Introduction

Welcome to this reissue of Max Hunter's seminal textbook on rocket science and engineering, *Thrust Into Space*, originally published by Holt, Rinehart and Winston in 1966.

"The book is written for the modern, technically-oriented high school student," Max wrote in his preface in 1966. It's been said that this book is unique not only in the topics it covers, but in the readable manner in which Max explains complicated physics with appropriate math - "comparatively simple expressions."

The original book is reprinted in its entirety, exactly as it was originally published. This introduction plus the foreword by Dr. J. D. Crouch, II appear before the original book begins.

Starting with page 225, you'll find 14 pages of new material including a list of major technical publications and policy papers, information on the Maxwell W. Hunter Foundation and *MaxwellHunter.com* which is a great resource for additional content on the career of Max Hunter.

Matthew S. Hunter

THRUST INTO SPACE

OFFICIAL 50TH ANNIVERSARY **REISSUE** **MAX HUNTER'S** INFLUENTIAL BOOK ON ROCKETRY & SPACEFLIGHT ENGINEERING

Foreword

Max Hunter was truly one of the legends of the American space program. I got to first know Max Hunter in 1986 long after the original publication of this book when I served as the military legislative assistant to Wyoming Senator Malcolm Wallop.

His passion, influence and leadership on many U.S. space-related projects and policies over five decades were unparalleled in the history of American space development.

A true American original with a rare combination of technical expertise and long-term vision, Max had the ability to clearly express his views both publicly and privately. He was one of our nation's preeminent modern visionaries.

At Douglas Aircraft Company, he was responsible for the aerodynamic design of Nike-Ajax and Hercules and other cold war missiles. Later, as Chief Missiles Design Engineer, Max was the principal designer of the legendary Thor IRBM, and, as Chief Engineer of Space Systems, for the engineering of all Douglas space efforts, including Thor's civilian counterpart, Delta, and the Saturn S-IV and S-IVB stages.

As part of the professional staff of both the Kennedy and Johnson National Aeronautics and Space Councils, he was first to recognize the effect of Jupiter's gravity on planetary probe vehicles and was instrumental in opening the outer solar system by supplementing rocket performance with planetary gravitational impulse.

At Lockheed Missiles & Space Company, he was responsible for early designs of the Space Shuttle, originated the concept of using large expendable tanks in shuttle design and led the Lockheed team through the design phase of Hubble Space Telescope.

In addition to *Thrust Into Space*, he left a rich legacy in the form of dozens of technical and policy papers, a list of which appears at the end of this book.

During the many meetings between me, Senator Wallop and Max, we would discuss policy issues, mostly regarding missile defense. After these meetings, the Senator would leave the two of us to lunch alone, resulting in some fascinating discussions. Senator Wallop would remind me, 'Max Hunter is a source of original thinking."

Max was a friend and mentor to me on technical issues, and shared with me his vision, passion and expertise about missile defense systems. His ideas years later greatly influenced my tenure as Deputy National Security Advisor, which brought about the deployment of a missile defense system in the United States.

Max always believed that maintaining the superiority of American technology would be the best method for the United States to not only defend itself but to maintain America's world economic leadership. That is one of the reasons that it is such a great honor to write this foreword for this re-publication of *Thrust Into Space*.

The ideas in this book were "written for the modern, technically-oriented high school student" in 1966 but are as relevant as ever. I hope its re-publication will help pave the way for and inspire more of America's youth to pursue an education in engineering – following the noble path of Max Hunter. - Dr. J. D. Crouch, II

Dr. Crouch holds a B.A., M.A. and Ph.D.in International Relations from the University of Southern California. He was the Principal Deputy Assistant Secretary of Defense for International Security Policy during the George H.W. Bush administration, was an Associate Professor of Defense and Strategic Studies at Missouri State University, served as U.S. Ambassador to Romania in 2004-2005 and was Assistant to President George W. Bush and Deputy National Security Advisor from March 2005 until June 2007. He currently is President and CEO of the United Service Organizations (USO).

Thrust into Space

Maxwell W. Hunter, II

Coordinating Editor:
James V. Bernardo, Director
Educational Programs and Services
National Aeronautics and Space Administration

 Holt, Rinehart and Winston, Inc., New York

Maxwell W. Hunter, II received an A.B. degree in Physics and Mathematics from Washington and Jefferson College in 1942, and an M.S. in Aeronautical Engineering from Massachusetts Institute of Technology in 1944. He joined Douglas Aircraft Company as a member of the Aerodynamic Performance Group, and then for eight years was in charge of the Missiles Aerodynamics Group which was responsible for the aerodynamic design of Nike-Ajax and Hercules, Sparrows, Honest John, and other missiles. In 1956 he was made Chief Missiles Design Engineer, responsible for the design of Thor, Nike-Zeus, and others, and in 1958 became Assistant Chief Engineer, Space Systems, responsible for all Douglas space efforts, including the Delta, Saturn S-IV stage, and others. In 1961 he was made Chief Engineer for Space Systems, and in 1962 joined the professional staff of the National Aeronautics and Space Council. In 1965 he assumed his present position as Special Assistant to the Vice President and General Manager of Research and Development at Lockheed Missile and Space Company.

Mr. Hunter is a Phi Beta Kappa, Tau Beta Pi, a Fellow of the American Institute of Aeronautics and Astronautics, and a senior member of the American Astronautical Society.

Original drawings: Versatron Corporation

Preface

The claim that propulsion is the key to space exploration has been repeated so often it has become trite. Even so, its importance is not always grasped. Many difficult techniques must be mastered in order to conquer space. It is important to spend time and effort learning to do them all well. Guidance and communication are examples. But it is inadequate propulsion—nothing else—which has limited the human race to one planet thus far. With excess propulsion, any guidance problem could be solved by carrying enough corrective thrust capability and any communication problem with enough power and equipment. All the sophisticated guidance and communication techniques in the universe, however, are of no help if the vehicle cannot carry a useful payload to space.

In the future, propulsion is undoubtedly the key to the magnitude of further space exploration and exploitation. Once such techniques as guidance and communication are mastered, miniaturized and routinized, they become relatively fixed-cost adjuncts of programs. Propulsion, however, determines the economic feasibility of space operations. Size of vehicles and fuel loads carried represent the fundamental price to be paid for space transportation. The question is whether the future of space exploration is merely to provide an expensive playground for select scientists and astronauts, as most people currently think, or whether it is to provide a vastly expanded domain for the entire human race, adding whole planets as the new worlds of the future. The answer lies strictly in the economics of space propulsion.

In this book, I have attempted to cover propulsion from the viewpoint of the systems architect, rather than the propulsion designer. Each chapter covers a certain velocity region. Each contains a discussion of basic flight mechanics of that region as an aid to understanding the appropriate propulsion systems.

Some knowledge of flight mechanics is essential to any real understanding of propulsion systems.

This book is written for the modern, technically-oriented high school student. Only comparatively simple expressions are utilized. Much of the massive calculations performed today are used to refine the last ounce of performance out of very complicated systems. This refinement is justified in design procedures, but use of the complicated calculations creates some risk that the user will lose sight of the fundamentals. Whether high school student or executive, basic decisions must be clearly related to the fundamentals in today's complicated technical world.

Today, we are engaged in materializing a two-thousand-year old dream of mankind. These dreams, and the restless, inquisitive drive of the human race to achieve its dreams, are the reasons we are going to space. I believe strongly that we should buckle down to the hard and spectacular job of engineering those dreams. That is why most of this book is devoted to future propulsion capabilities, not past propulsion history.

Valuable discussions, sometimes heated, with innumerable colleagues, throughout the years, contributed in many ways to this book. The advice and encouragement of Dr. Edward C. Welsh, Executive Secretary of the National Aeronautics and Space Council, was essential. Special thanks is due to Dr. Charles S. Sheldon II of the staff of the National Aeronautics and Space Council, Robert F. Trapp of the National Aeronautics and Space Administration, and Joseph M. Tschirgi of Bellcomm, Inc., for refinement of the text. If anything is unique in this book, however, it must be the assistance of a leading lady of the American theatre, Irene Manning, my wife. Rarely has so much charm and talent been devoted to the typing of a manuscript and the enthusiastic clarification of a text. Her help was invaluable.

M. W. H.

Contents

		Page
1	Rocket Fundamentals	7
2	Artillery Rockets	28
3	Orbital and Global Rockets	52
4	Lunar and Early Interplanetary Rockets	78
5	Solar System Spaceships	140
6	Interstellar Ships	188
7	Outlook	208

List of Symbols	213
Bibliography	215
Glossary	216
Index	222

1

Rocket Fundamentals

Introduction

The thrust of the human race into space is primarily a story of man's ability to achieve higher velocities. About the highest velocity a man can attain by running is 32 feet per second (fps) (60 miles per hour = 88 feet per second). The highest velocity a human can attain, relying primarily on his muscles, is to use a bow to launch an arrow to achieve about 350 feet per second. A typical high velocity gun, such as an anti-aircraft gun or the German World War I "Paris gun," achieves about ten times bow and arrow velocity, or 3500 feet per second. An intercontinental ballistic missile (ICBM) normally attains a maximum velocity of about 23,000 feet per second. An orbit around the earth requires 26,000 feet per second. To completely escape from earth requires 36,700 feet per second, or about ten times the velocity of the Paris gun. The velocity required to launch directly from earth and escape from the solar system is about 54,600 feet per second.

For complex missions where velocity is added or taken away in several increments, rocket designers normally speak in terms of the total of these increments. The total velocity requirement for going to the moon, landing, and returning to earth with atmospheric braking on return is about 54,000 feet per second, almost 50 per cent greater than earth escape velocity. Although such velocities and some much higher will be discussed, it is well to remember that earth escape velocity was first achieved only six years ago on January 2nd, 1959, by the Russian Lunik I and that it took at least 700 years of human ingenuity before rockets achieved that velocity.

Normal rocket engines operate by creating a high pressure and temperature within a chamber, then converting the thermal energy so released into a useful force by expanding the gases rearward through a nozzle at high velocity. A continual supply of high temperature and pressure gas is created by burning rocket fuel in the chamber. As the gas goes through the nozzle, its pressure and temperature decrease as thermal energy is converted to velocity, and the thrust on the vehicle is basically the recoil from the exhausting of the gas.

All rockets operate on this recoil from their exhaust. Rockets using the process described are termed "thermal rockets," regardless of whether their source of energy is chemical or nuclear. Rockets which use electrical means to generate a high velocity exhaust will be discussed in Chapter 5.

Guns

To understand rocket fundamentals, it is useful to examine a gun in some detail. When a gun is fired, the powder releases chemical energy which creates a hot high pressure gas which then expands with decreasing pressure and temperature while pushing the bullet down the barrel ahead of it. This process occurs very rapidly, in just over one millisecond (one thousandth of a second) for a modern high velocity rifle, producing 4000 feet per second muzzle velocity in a 26-inch barrel.

Newton's Third Law of Motion states that when two bodies react, the action and reaction forces are equal and opposite. This is the law of conservation of linear momentum. Expressed mathematically, when the bullet leaves the muzzle, the momentum of the gun will be equal and opposite to the momentum of the bullet. Momentum is the product of mass times velocity, and hence we have the following equation:

$$m_G V_G = m_B V_B \qquad (1\text{-}1)$$

where m_G = mass of gun in slugs; m_B = mass of bullet in slugs; V_B = muzzle velocity of bullet in feet per second; and V_G = recoil velocity of gun in feet per second.

As the gas is forcing gun and bullet apart, the bullet pushes on the gun and the gun on the bullet. The recoil on the gun has nothing to do with the fact that the bullet pushes through air after it leaves the muzzle.

Force and Energy

Mass is simply the quantity of matter. Force is related to mass by Newton's Second Law of Motion which is (for constant mass):

$$F = ma \qquad (1\text{-}2)$$

where a = acceleration in feet per second^2; F = force in pounds; and m = mass in slugs.

Equation 1-2 indicates a convenient means of measuring mass—namely, by weighing a quantity of matter under the acceleration of gravity. Thus, a pound of mass is defined as the quantity of matter which weighs (exerts a force of) one pound at the earth's surface. Since the acceleration of gravity varies somewhat over the earth's surface (it is about 0.5 per cent lower at the equator than at the poles), a standard value of 32.174 feet per second^2 has been agreed upon for the purpose of standard force and weight measurement. When measuring weight specifically, Equation 1-2 becomes:

$$w = mg_o \qquad (1\text{-}3)$$

where w = weight in pounds; m = mass in slugs; and g_o = 32.174 feet per second^2.

In other words, one slug weighs (exerts a force of) 32.174 pounds under standard gravity. The standard weight will be used throughout this book as a measure of mass since it is a familiar term. One normally speaks of a Thor rocket as weighing 100,000 pounds, not as having a mass of 3,100 slugs.

To avoid confusion, the symbol w will represent weight of mass in pounds. The mass in slugs will always be represented by the symbol m. The same quantity of mass would weigh different amounts if weighed on different planets, but we will use only earth standard weight.

Energy is defined as the ability to do work. Work is measured by the product of force and distance when the force is moving a mass against a resistance. Hence, both work and energy are measured in foot-pounds. Work must be performed on a body in order to create velocity, and a moving body possesses energy by virtue of its motion equal to the work performed on it. This is called its kinetic energy, and is given by the expression:

$$KE = \frac{1}{2} mV^2 = \frac{1}{2} \frac{w}{g_0} V^2 \qquad (1\text{-}4)$$

where KE = kinetic energy in foot-pounds.

The ratio of kinetic energies of gun to bullet may be obtained from Equations 1-1 and 1-4 as:

$$\frac{KE_G}{KE_B} = \frac{V_G}{V_B} = \frac{w_B}{w_G} \qquad (1\text{-}5)$$

where KE_G = kinetic energy of gun in foot-pounds; and KE_B = kinetic energy of bullet in foot-pounds.

Although momentum may be equally divided between two bodies, energy need not be. More energy goes into the bullet than the gun. Since guns weigh on the order of 500–1000 times the bullet they fire, almost all the energy goes into the bullet. Were this not so, either end of the gun would be equally destructive.

Efficiency

The efficiency of the gun can be determined by comparing the energy of the bullet with the energy release of the amount of powder burned. Heat energy releases are commonly measured in the units of BTU per pound where BTU stands for British Thermal Unit. They are also frequently quoted in gram-calories per gram. (One gram-calorie per gram = 1.8 BTU per pound). Energy can be converted from one form to another, and the mechanical energy equivalent of heat energy has been determined to be 778 foot-pounds per BTU. The energy release of modern smokeless gunpowder is about 1250 gram-calories per gram, so that it is equivalent to 1,750,000 foot-pounds per pound, or 250 foot-pounds per grain. (It is common to use grain as a measure of mass in small arms work for both powder and bullets. 7000 grains = one pound.)

Typically, the use of about 40 per cent of the bullet's weight in powder will generate about 3000 feet per second muzzle velocity which is approximately 20 foot-pounds of energy per grain of bullet (see Equation 1-4). Guns, at least of the hand-held variety, are typically about 20 per cent efficient. The energy not used in propelling the bullet is dissipated in friction between bullet and barrel, used to move air from the barrel, or is carried away by the powder gases, both as kinetic energy of the gases and as thermal energy contained in the hot gases.

Power

When considering rockets or any other propulsion device, both the energy consumed and the power involved are of interest. Power is defined as the time rate at which work is done, or energy released. Thus, energy is measured in foot-pounds and power in foot-pounds per second. If the energy is released fast enough, the power can be very high, even though the total energy release is low. For a rifle which burns 40 grains of powder to fire a 100 grain bullet at 3000 feet per second, the total energy release within the chamber is of the order of 10,000 foot-pounds. The time of travel down the barrel will be about 1.44 milliseconds. Thus, the rate of energy release is 6.9 million foot-pounds per second. Since one horsepower is 550 foot-pounds per second, this is 12,500 horsepower. A comparable number for a very powerful hunting rifle is 53,000 horsepower. We normally measure horsepower in terms of useful work, not energy release. With 20 per cent efficiency, the hunting rifle delivers 10,600 horsepower. It is perfectly correct to say that such a gun, although easily carried, has over three times the horsepower of a modern railroad locomotive weighing over 150 tons. It is necessary to keep power and energy clearly separated in one's mind.

Guns as Rockets

Consider building a rocket which was simply a gun—which ejected only one particle rearward. Equation 1-1 may be rewritten in terms of the relative velocity between the two masses as follows:

$$\Delta V = v_e \left(1 - \frac{w_F}{w_I} \right) \qquad (1\text{-}6)$$

where ΔV = velocity achieved by final mass in feet per second; v_e = relative velocity of final and ejected masses in feet per second; w_I = initial (total) weight in pounds; and w_F = final weight in pounds.

The maximum velocity is limited to the relative velocity of ejection and this is only achievable in the extreme of zero final mass.

Ejecting more than one mass would be beneficial. For instance, if two masses were ejected of such size that half the remaining mass were ejected each time, then the velocity increment achieved each time would be equal to one-half the ejection velocity, and the ratio of initial to final mass required each time would be two. Hence, the final velocity achieved would be equal to ejection velocity, but the initial to final mass ratio would be only four rather than infinite.

By using Equation 1-6 repetitively for any discreet number of masses, it can be shown that the most efficient process is a continuous ejection of a large number of very small masses. All rockets make use of such a continuous stream of gas rather than discreet particles. The simple momentum and energy relations of the previous equations must be replaced by somewhat more complicated continuous flow derivations. Most of the equations presented will not be rigorously derived, but the method of derivation will be indicated.

Rocket Engines

The thrust due to the exhaust velocity of the gases as they leave the nozzle of a rocket is not the total force on the rocket. Since a gas supply is continually being created in the chamber and continually leaving the rocket system, the pressure on the surface through which this gas leaves must also be considered. By making use of Newton's Second Law and equating the momentum of a rocket vehicle flying through an atmosphere with the momentum of its exhaust jet and the surrounding atmospheric pressures, the following equation for rocket thrust is obtained (see Figure 1-1):

$$T = \frac{\dot{w}}{g_0} v_e + (p_e - p_{at})A_e \qquad (1\text{-}7)$$

where T = thrust (force) of rocket in pounds; $\dot{w}$ = propellant flow rate in pounds per second; v_e = exhaust velocity in feet per second; p_e = exhaust gas pressure at nozzle exit in pounds per square inch (psi); p_{at} = local atmospheric pressure in pounds per square inch; and A_e = nozzle exit area in square inches.

The thrust consists of two terms. The first term, called momentum thrust, is the product of the propellant flow rate and its exhaust velocity. The second term, the pressure thrust, is the nozzle exit area multiplied by the difference between atmospheric pressure and nozzle exit pressure.

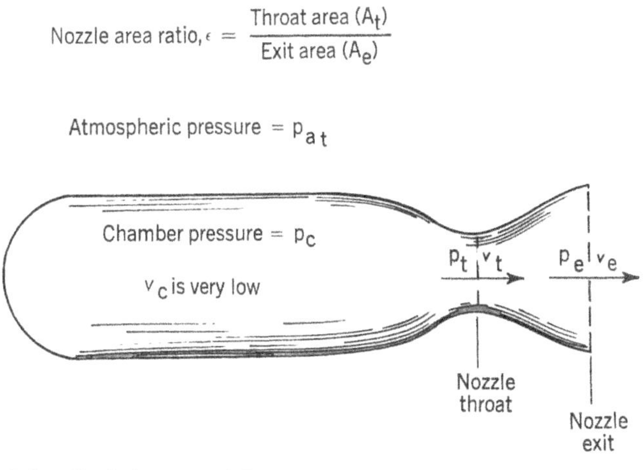

$$\text{Nozzle area ratio, } \epsilon = \frac{\text{Throat area } (A_t)}{\text{Exit area } (A_e)}$$

Atmospheric pressure $= p_{at}$

Fig. 1-1.   Rocket nomenclature.

Fuel Consumption

A large thrust can be obtained either by ejecting a large amount of propellant or by ejecting a small amount at a high velocity. It is preferable to have rockets of low propellant consumption. A measure of propellant consumption is the thrust obtained for a given amount of propellant used per second.

$$I_{sp} = \frac{T}{\dot{w}} \tag{1-8}$$

The quantity I_{sp} is widely used by rocket engineers and is called the specific impulse of the engine. It has the dimensions of time since it is thrust in pounds divided by propellant consumption in pounds per second. It is the time for which one pound of propellant could produce one pound of thrust. Effective exhaust velocity is a fictitious velocity which includes both momentum and pressure thrust, and is defined as:

$$v_{ef} = g_o I_{sp} \tag{1-9}$$

Power

The power expended in the rocket exhaust is the kinetic energy per second of the jet which is given by the expression (see Equation 1-4):

$$P_e = \frac{\dot{w} v_e^2}{2g_o} \tag{1-10}$$

where P_e = power in foot-pounds per second.

A convenient approximation to Equation 1-10 is to replace the actual exhaust velocity (v_e) with the effective exhaust velocity (v_{ef}) so that:

$$P_{ef} = \frac{\dot{w} v_{ef}^2}{2g_o} = \frac{T v_{ef}}{2} = \frac{g_o I_{sp} T}{2} \tag{1-11}$$

where P_{ef} = effective power in foot-pounds per second.

The effective power is frequently a good approximation to actual power, since the pressure thrust term of Equation 1-7 is small in many instances.

These simple equations show clearly some basic points about rocket propulsion. If low fuel consumption is desired, Equation 1-7 shows that high exhaust velocity is required. Equation 1-11 shows that for a given thrust, high exhaust velocity requires increased power in the exhaust which results in increased energy release within the rocket chamber. Much of this book will discuss the continual fight for higher performance rocket engines which means the controlled release of increasingly larger amounts of energy. Periodically, reference will be made to these simple expressions as a means of grasping the magnitude of the forces and energies involved.

Internal Energy Release

If it were possible to convert all thermal energy in the combustion chamber to nozzle exit velocity, then the expression for exit velocity would be:

$$v_e = \sqrt{2g_oJh} = 224\sqrt{h} \qquad (1\text{-}12)$$

where h = enthalpy per unit weight in BTU per pound; and J = mechanical equivalent of heat (778 foot-pounds per BTU).

Enthalpy is the term applied to the total heat released by the combustion process.

Complicated engineering calculations are required to go from Equations 1-10 and 1-12 to the actual energy release within the rocket chamber. Efficiency of the combustion process, heat lost through the chamber walls, fluid friction losses and flow angularities in the nozzle, and other phenomena must be considered. For efficient rockets, all of these effects amount to less than five per cent of the total energy except for the amount of energy lost due to the temperature of the exhaust jet. Only the kinetic energy of the exhaust jet is useful. Any thermal energy remaining in the hot jet represents a penalty exactly analogous to the thermal energy lost in the hot exhaust of any closed cycle combustion engine, or that lost in the hot gas from the muzzle of a gun.

A nozzle accelerates flow by expanding it so that the gas

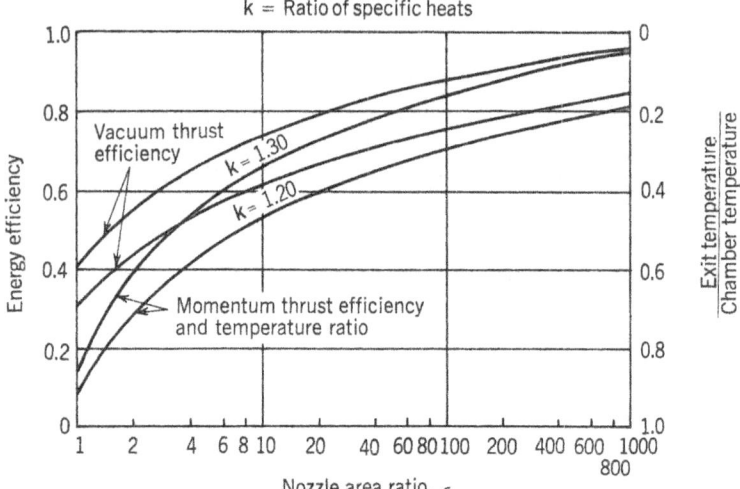

Fig. 1-2.   Rocket energy efficiency.

temperature decreases. The nozzle expansion is measured by the ratio of nozzle exit area to throat area, called the expansion ratio (ϵ). A nozzle of infinite expansion ratio, operating in a vacuum, should produce complete conversion of the thermal to kinetic energy, and Equations 1-10 and 1-12 would give the internal energy release to within five per cent. Such large nozzles are impractical. Furthermore, there is a point at which the exhaust gas cools down to where it liquefies and the whole nozzle expansion process breaks down.

Figure 1-2 shows both the momentum thrust and the total vacuum thrust of nozzles as a function of the nozzle-exit-to-throat-area ratio and the ratio of specific heats (k) of the propellants. The value of k varies between 1.2 and 1.3 for almost all rocket propellants. It is important to note that a rocket engine in a vacuum can easily convert over 80 per cent of propellant energy into useful thrust. This number is usually about 20 per cent for closed cycle thermal combustion engines and guns. *If rockets did not have to carry all propellant aboard, they would be at least three times as efficient as other thermal propulsive devices.*

The combustion temperature of chemical propellants tends to be limited, not only due to limited energy release, but also because above about 4000 degrees Fahrenheit (°F), increasing amounts of energy go into breaking apart the gas molecules (a phenomenon called dissociation) rather than raising their temperature. Temperature is a measure of the average kinetic energy of the gas. Applying Equation 1-4 to an individual molecule:

$$T_{co} \sim \frac{MV^2}{2} \qquad (1\text{-}13)$$

where V = velocity of molecule; T_{co} = temperature of combustion; and M = molecular weight. Hence:

$$V \sim \sqrt{\frac{2T_{co}}{M}} \qquad (1\text{-}14)$$

This molecular velocity is a measure of exhaust velocity. Hence, higher exhaust velocity is achieved for a given combustion temperature with a lower molecular weight exhaust gas. As in the case of gun and bullet, where energy was not equally divided between the two although momentum was, the momentum of different weight molecules varies even when their kinetic energies are the same.

Any energy which goes into dissociation of the combustion gases in the chamber is not useful for accelerating the gases. In many cases, recombination of the molecules occurs as the gases cool while flowing through the nozzle and the energy is released there as thermal energy. When this happens, the efficiency as given in Figure 1-2 no longer applies. If no change in composition of the combustion gases occurs in the nozzle, the flow is said to be in frozen equilibrium. If recombination occurs so rapidly that the composition is that which would occur normally at the temperature and pressure of the nozzle, the flow is said to be in shifting equilibrium. In actuality, something between these two cases occurs.

Since both dissociation and recombination rates are functions of pressure and temperature, very complicated thermo-

chemical calculations are required to estimate rocket performance. It is more difficult than measuring the energy release of
gasoline burning or gunpowder exploding at a constant pressure. Extensive computer calculations have been performed
for many propellants for both frozen and shifting equilibrium.
Results are usually presented as specific impulse for various
combustion pressures and nozzle area ratios. It is necessary to
use specific impulse for performance comparison rather than
energy release due to the strong effect of molecular weight on
specific impulse as well as the general complication of the
chemical phenomenon. We shall use specific impulse or effective exhaust velocity, with the efficiencies shown in Figure 1-2
only as an approximate guide to the magnitude of basic
processes.

Atmospheric Pressure Effect

Rocket performance in the atmosphere is not as high as in
the vacuum of space. This is because the exhaust jet must of
necessity displace a portion of the atmosphere, as shown by
the pressure thrust term in Equation 1-7. Figure 1-3 illustrates
nozzle flow characteristics as a function of external pressure.
At high altitudes where external pressure is lower than nozzle
exit pressure, the nozzle is said to be underexpanded, and the
flow aft of the nozzle expands rapidly. If the external pressure
is substantially higher than exit pressure (about 2.5 times),
then the flow in the nozzle separates. This is beneficial from a
thrust viewpoint, since if the nozzle were flowing full at the
lower pressures, the negative pressure thrust term would be
greater. Although separation increases thrust, it is usually difficult to predict accurately, and sometimes results in oscillating
flow conditions and excessive vibration.

Figure 1-4 shows the variation of rocket thrust with external
pressure for k = 1.30. The line labelled "maximum thrust at
given pressure" represents optimum expansion, or the maximum
thrust possible for a given chamber/external pressure ratio.
The same nozzle will give a higher thrust at higher pressure

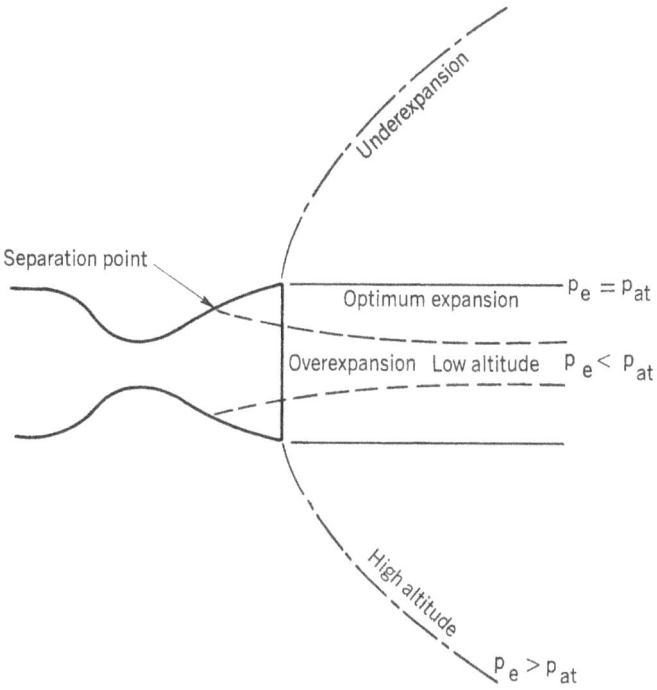

Fig. 1-3. Nozzle altitude effect.

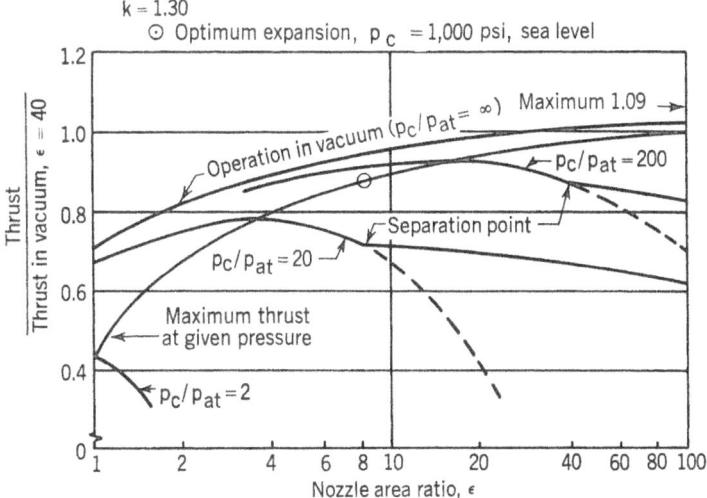

Fig. 1-4. Nozzle altitude performance.

ratio. Figure 1-4 is drawn as a ratio of thrust to that available with expansion ratio of 40 in a vacuum. Rocket engineering practice for decades has used optimum expansion with 1000 pounds per square inch chamber pressure at sea level as a standard for propellant comparison. Since this book will discuss mostly space performance, a standard vacuum performance is also pertinent. The value with expansion ratio of 40 will be used as such. The difference between sea level and vacuum standards is about 15 per cent. Figure 1-4 may be used to convert to other expansion ratios and altitudes with an accuracy of a few per cent for most propellants.

Pump Power

In liquid rocket engines, pumps are frequently used to raise the pressure of the propellants from tank pressure to combustion chamber pressure. Pump power is a function of pressure rise, fluid density, and flow rate. The power required for a pump of perfect efficiency is given by the expression:

$$P_{pu} = \frac{144 \Delta p \dot{w}}{\rho} \qquad (1\text{-}15)$$

where P_{pu} = pump power in foot-pounds per second; Δp = pressure rise in pounds per square inch; and ρ = fluid density in pounds per cubic foot.

In general, two propellant fluids will have different densities, and tank pressures may also be different. Hence, Equation 1-15 must be applied to each pump separately in bipropellant engines. If tank pressures and fluid densities are averaged, then the pump power for both propellants may be approximated with the aid of Equation 1-8 as:

$$P_{pu} = \frac{144 \Delta p T}{\rho I_{sp}} \qquad (1\text{-}16)$$

Since pump efficiencies usually are between 0.45 and 0.65, about twice the power as given by these equations must be used to drive the pumps.

The Rocket Equation

The velocity that a rocket achieves is a function of many things. If we ignore extraneous forces, such as gravity and atmospheric drag, the velocity achieved is a function of the amount of propellant carried and its effective exhaust velocity. The equation relating these quantities is:

$$\Delta V = v_{ef}\ln\frac{w_I}{w_F} = g_o I_{sp}\ln\frac{w_I}{w_F} \qquad (1\text{-}17)$$

where ΔV = velocity change in feet per second; w_I = initial weight of rocket in pounds; and w_F = final weight of rocket in pounds. ΔV is known as the impulsive velocity to differentiate it from the actual velocity change including drag and gravity effects. This is the classical rocket equation. It is plotted as the top curve in Figure 1-5. Equation 1-6 is an approximation to it at very low velocity increments.

Practical rockets unfortunately require structure to contain the propellants. The weight of the thrust chambers, nozzles, and other equipment must also be considered. Consequently, high velocities can only be obtained by discarding part of the weight along the way. This is known as staging. Most rockets currently use several stages, since practical stage empty weights will not permit the high velocities desired. Theoretically, it would be nice to discard weight continuously (infinite staging). The rocket would then behave simply as if it had a greater propellant consumption by the amount of weight discarded, and the rocket equation would become:

$$\Delta V = \lambda' v_{ef}\ln\frac{w_I}{w_F} \qquad (1\text{-}18)$$

where λ' is defined as the propellant weight divided by the total propulsion system weight (propellant, engines, tanks, controls, etc.).

$\lambda' = 1.00$ means zero rocket structural weight and represents an ultimate limit. Modern rockets possess λ' between 0.80 and 0.97. Rockets which existed at the start of the Twentieth Century possessed λ's of about 0.25. The effect of infinite staging is shown in Figure 1-5.

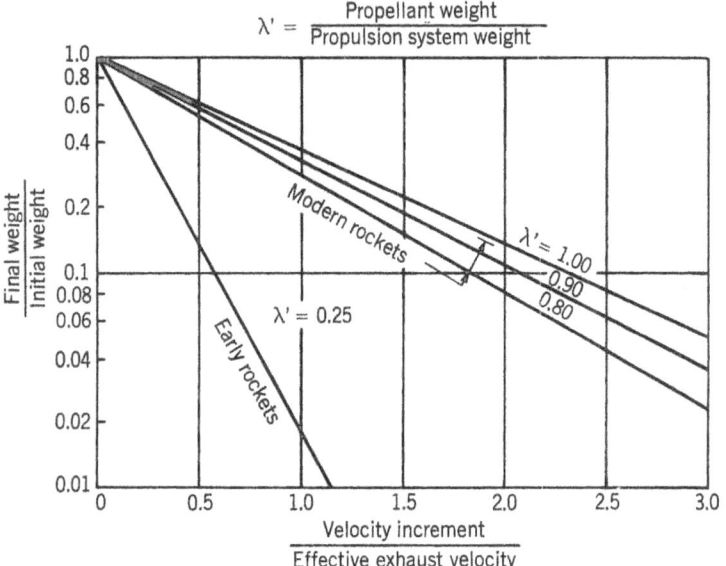

Fig. 1-5.   The rocket equation.

Much can be learned by understanding the classical rocket equation. Figure 1-5 shows that rockets can be made to go to any desired velocity, regardless of their own exhaust velocity, as long as the final weight is small enough compared to the initial weight. This figure can be extrapolated to any desired velocity by extending the curves shown. A convenient rule to remember is that for a given final weight, the rocket weight must increase by a factor of 10 for every velocity increase of about twice the exhaust velocity for modern rockets with staging approximating the ultimate. Penalties in rocket design may occur up to ridiculous extremes, however, if one pursues high velocity rockets too recklessly. An example worked out 50 years ago is pertinent:

When Goddard first started his experiments, he measured the exhaust velocities of the rockets in use at that time as about 1000 feet per second (I_{sp} = 31 seconds). These rockets carried only about 25 per cent of their weight as fuel. Figure 1-5 shows that even with infinite staging, the initial rocket weight would be 10 times the final weight at a velocity of about 580 feet per second. If one were to try to generate a velocity equivalent to a modern Thor intermediate range ballistic missile by staging such rockets, a typical velocity of 14,500 feet per second would be 25 times 580 feet per second. The rocket initial weight would be 10 multiplied by itself 25 times or 10^{25}. The weight of the earth is 1.32×10^{25} pounds. In this case, it would require the entire weight of the earth to put one pound up to IRBM velocity.

Goddard's first report in 1919 revealed excellent experimental work and superb imagination. It also revealed that when Goddard worked out a similar example compared to the weight of the earth, an error was made of a factor of about 27,000,000 in weight of the earth. This may be called a sizable error. Whether misprint or mistake, it was still uncorrected and apparently unnoticed when the report was reprinted by the American Rocket Society in 1946.

With modern rocket structures, where the weight ratio with infinite staging is 10 at twice the exhaust velocity, a rocket the weight of earth would launch one pound to 50 times exhaust velocity. Thus, even with modern structures, if only the 1000 foot per second exhaust velocity of the early rockets were available, the entire weight of the earth would not be enough to put one pound through the 54,000 foot per second lunar and return mission.

Useful Load

A useful form of Figure 1-5 can be obtained for single stage rockets by separating the weight of useful load carried from the weight of engines and structures necessary for propulsion. In this case, useful load (w_{UL}) is defined as everything including structure above the propulsion unit. Hence:

$$w_I = w_{UL} + \frac{w_{pr}}{\lambda'} \tag{1-19}$$

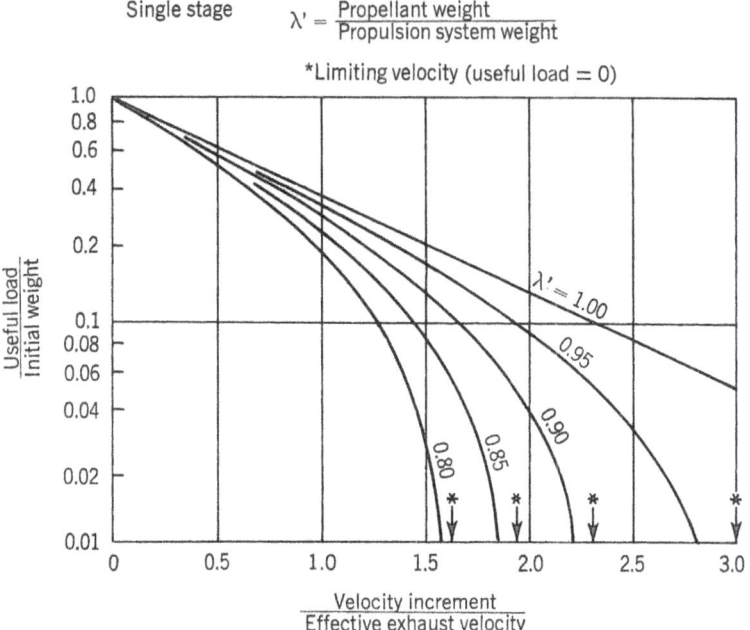

Fig. 1-6. The rocket equation.

where w_{pr} = weight of propellant in pounds; and w_{UL} = weight of useful load in pounds. Figure 1-6 shows the rocket equation in terms of useful load.

The curves of $\lambda' = 1.0$ in Figures 1-5 and 1-6 are identical. The limit on maximum velocity for a given λ' with one stage occurs where the weight of engine and structure equals the final weight required. Figures 1-5 and 1-6 may be compared to give an indication when staging is necessary to prevent excessive weight penalties. In practical multi-stage rockets, the performance is calculated by applying Figure 1-6 to each stage, and combining the total, assuming upper stages to be the useful load of lower stages.

Energy Efficiency

The efficiency of a rocket in performing useful work is given by the ratio of kinetic energy imparted to the useful load to kinetic energy expended in the exhaust. This ratio automatically includes the penalty for carrying propellant along. Under the assumption of zero initial kinetic energy, the final kinetic energy of the useful load is given by:

$$KE_{UL} = \frac{1}{2} \frac{w_{UL}}{g_o} \Delta V^2 \qquad (1\text{-}20)$$

and the total energy expended by the exhaust is given by:

$$KE_{ef} = \frac{1}{2} \frac{w_{pr}}{g_o} v_{ef}^2 \qquad (1\text{-}21)$$

The ratio of these terms is plotted in Figure 1-7, as the external energy efficiency. This figure shows that for good external energy efficiency the design velocity increment should be close to the effective exhaust velocity, with the actual optimum value a function of the stage λ'. The most important point to be learned from Figure 1-7 is that rocket external

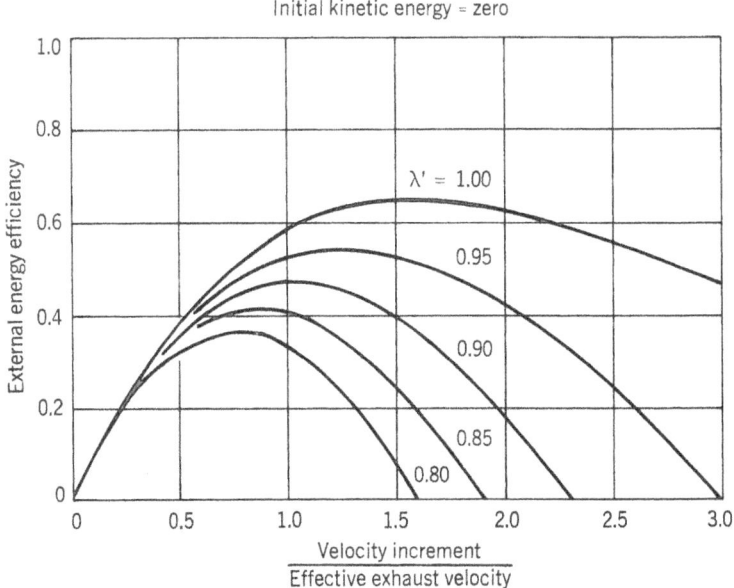

Fig. 1-7.   External energy efficiency.

energy efficiencies can easily be 40 or 50 per cent over a wide range of design velocities, as long as the velocity increment is somewhere around the effective exhaust velocity. Tsiolkovskiy showed this clearly in 1903. When this roughly 50 per cent external energy efficiency is combined with roughly 80 per cent ratio of internal energy to effective exhaust energy previously shown (Figure 1-2), the overall energy efficiency of a rocket can easily be 40 per cent. *Rockets can be almost twice as efficient as most internal combustion engines, even when the penalty of carrying all fuel aboard is included.*

Effect of Initial Velocity

The effect of initial velocity not equal to zero can also be obtained. To do this, the energy of the final mass must be compared to the energy expended in the exhaust plus the initial kinetic energy of the propellant. The kinetic energy increase of the useful load is:

$$KE_{UL} = \frac{1}{2}\frac{w_{UL}}{g_0}(V_F{}^2 - V_I{}^2) = \frac{1}{2}\frac{w_{UL}}{g_0}(\Delta V^2 + 2\Delta VV_I)$$

$$(1\text{-}22)$$

where V_I = initial velocity in feet per second; V_F = final velocity in feet per second; and ΔV = velocity increment of rocket in feet per second. The total energy expended in the jet includes the initial kinetic energy of the fuel and is given by:

$$KE_{ef} = \frac{1}{2}\frac{w_{pr}}{g_0}(v_{et}{}^2 + V_I{}^2) \qquad (1\text{-}23)$$

Energy efficiency in this case is shown in Figure 1-8 for $\lambda' = 1.00$.

Figure 1-8 shows that rockets which already have been given some initial kinetic energy can be extremely efficient in converting this energy to useful propulsive capability, as long as the exhaust velocity is of roughly the same magnitude as the other velocities involved. It is convenient to visualize rockets as able to generate high velocities because of their kinetic energy conversion capability. The later stages are able

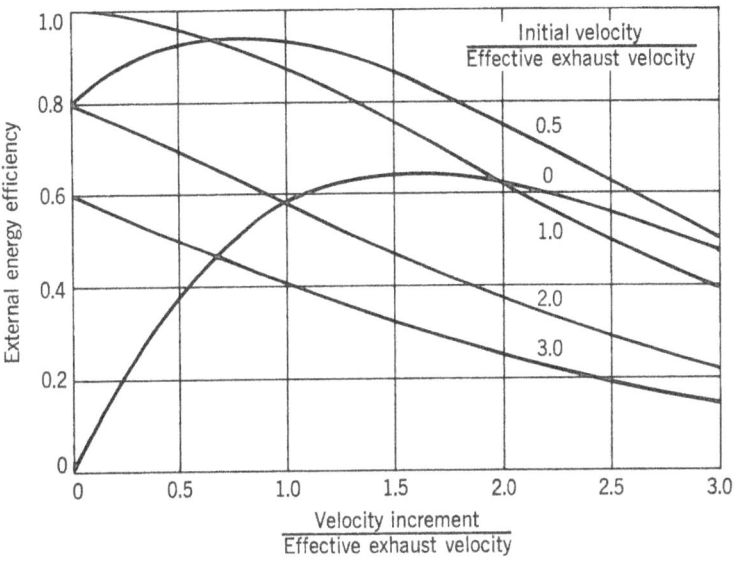

Fig. 1-8.   External energy efficiency.

to convert the kinetic energy already imparted to their propellant to further useful work.

Almost all propulsion systems obtain either fuel or oxidizer from the medium through which the vehicle moves. Since rockets must carry both aboard, the big problem in rocket design to date has been the weight of propellants which must be carried rather than the energy consumed. Low propellant weight and efficient energy utilization are contradictory requirements since low propellant weight requires high exhaust velocity, but that, in turn, requires higher internal energy release. Hence, energy efficiency has been mostly of theoretical interest to date, and overall weight gains made by expending extra energy have been mostly beneficial.

In the future, as higher performance rocket engines are able to remove the weight criticality and operate more like normal transportation devices, energy utilization will become important. In the long run, it is the price of energy which is fundamental.

2

Artillery Rockets
(Velocity up to 5,000 Feet Per Second)

Ballistics

The earliest use of rockets involved very low performance compared with what we now believe possible. These uses had nothing to do with space flight. They involved terrestrial applications of the same class of performance as conventional artillery. Velocity requirements under these conditions can be derived with simple assumptions. If the earth is assumed to be flat, and atmospheric drag is neglected, the following equations may be derived from Newton's Laws of Motion:

$$s = V_h t_t \qquad t_t = \frac{2V_v}{g} \qquad h = \frac{V_v^2}{2g} \qquad (2\text{-}1)$$

where V_v = vertical velocity in feet per second; V_h = horizontal velocity in feet per second; t_t = flight time in seconds; s = horizontal range in feet; h = altitude in feet; and g = acceleration of gravity in feet per second^2.

Maximum range is obtained for a given velocity by firing both upward and forward at the optimum combination of flight duration and horizontal velocity. The optimum angle is 45 degrees from the horizontal. The resulting relationship between range and velocity is:

$$s = \frac{V^2}{g} \qquad (2\text{-}2)$$

Equations 2-1 and 2-2 show that the maximum altitude achieved by firing vertically is exactly one-half the maximum range possible.

Appreciable distance may be covered during the motor burning time of large rockets. A V-2, with a burnout velocity of 5060 feet per second, would have a range according to Equation 2-2 of 150 miles. This is only the range from burnout to the point down range where the rocket has returned to burnout altitude. To this must be added the distance covered prior to burnout, about 14 miles, and a similar distance at the end of flight since the rocket is returning to earth at the approximate 45 degree angle at burnout.

When calculating accurately the impact points of rockets or guns, even of ranges of only a few miles, it is necessary to include second order corrections for the curvature and rotation of the earth. Only the simplest form of the ballistics equations, however, are presented here. Such expressions can be a great aid in understanding the need for different types of rocket propulsion for various missions. Artillery rockets are defined as those with velocities up to 5000 feet per second. This includes performance up to the German V-2 rocket, which had a range of almost 200 miles, or more than twice that of the longest range gun ever used.

Energy

The concept of the kinetic energy a body possesses due to its motion is discussed in Chapter 1 and the expression for kinetic energy is given in Equation 1-4.

A body may also possess energy by virtue of its position with respect to another body, if a force field is involved. This type of energy is called potential energy. The force field of interest here is the gravitational force field. If a projectile is dropped from a height, it is accelerated by gravity and has a certain velocity, or kinetic energy, when it hits the surface. If there are no energy dissipating forces present (such as atmospheric drag), the total of potential and kinetic energy must remain constant, although one may be changed into the other. For the flat earth case under discussion, potential energy is given by:

$$PE = mgh = wh \qquad (2\text{-}3)$$

where PE = potential energy in foot-pounds.

Thus, as an object moves according to Equation 2-1, its kinetic energy decreases as it rises away from earth and the gravity field decreases the vertical velocity component. At peak altitude, the potential energy is greatest, but it decreases again to zero at zero altitude and is converted back to the original kinetic energy. The sum of potential and kinetic energy remains constant.

The fact that kinetic energy increases as the square of the velocity affects the conversion of energy requirements to rocket performance. The kinetic energy created by a velocity input is strongly dependent on the initial velocity, as shown by Equation 1-22. The ratio of kinetic energy increase to that if the initial velocity were zero is given by:

$$\frac{\Delta KE}{\Delta KE_0} = 1 + 2 \frac{V_I}{\Delta V} \tag{2-4}$$

The second term can easily be larger than one. Hence, although the velocity increment of a rocket as shown by the rocket equation is a function only of the weight ratio and exhaust velocity, the kinetic energy increase also depends on the initial velocity. If kinetic energy is converted to potential energy, subsequent velocity inputs will be at lower initial velocity and less effective in producing kinetic energy.

A general conclusion is that all velocity inputs should be made at the highest kinetic energy possible, which means the lowest potential energy which, in turn, means the lowest altitude. The energy expressions are useful in understanding rockets, but they must be converted to total velocities in order to be applied to the rocket equation. We shall deal in both energies and velocities as appropriate.

The ballistic equations represent only one portion of a rocket's travel. In general, travel can be broken down into three parts: starting, mid-course, and stopping. Two major dissipating forces—gravity and atmospheric drag—affect all these regions differently for various forms of travel. It is useful to examine these effects, not only for rockets and guns, but for land, sea, and air transportation as well.

Atmospheric Drag

In artillery projectile flight, atmospheric drag causes a significant deviation from the simple formulae 2-1 and 2-2. The projectile is slowed substantially in the dense lower atmosphere. Equation 2-2 gives about 41 miles as the range corresponding to a velocity of 2650 feet per second, but the maximum range of a 16-inch naval gun of that muzzle velocity is under 26 miles. The projectile loses over 500 feet per second velocity in the first six miles of travel. A rifle of 3000 feet per second muzzle velocity loses 500 feet per second in only about 200 yards of travel.

Some short range rockets generate their velocity quickly in the low atmosphere and are affected by atmospheric drag in much the same manner as artillery projectiles. Long range rockets climb relatively slowly out of the atmosphere and burnout usually occurs so high they are relatively unaffected by the atmosphere until re-entry. Even while climbing through the atmosphere, the drag penalty is lower than one might think from the gun numbers quoted above.

The force due to drag is a complicated function of body shape and surface area, flight velocity, and various atmospheric parameters. The drag/weight ratio determines the deceleration. The large difference between heavy artillery projectiles and small arms is due to the ability to place a greater weight per surface area in the larger projectiles.

The drag force goes up rapidly with increasing velocity, but also decreases rapidly at high altitudes. This effect is noticeable in the drag loss of long range artillery. At mid to long ranges, the drag loss of the 16-inch projectile is about 1000 feet per second, and actually is slightly less at maximum range. It would be over 2000 feet per second if the loss rate of the first six miles applied throughout. This is due to the long range shots being fired so high that they travel mostly in the high atmosphere.

Long range rockets not only travel at even higher altitudes than guns, but climb through the atmosphere at a slower rate, avoiding high drag peaks which a gun projectile experiences close to the muzzle. Typical drag losses for a ballistic missile are from 1500 to 2000 feet per second. This value applies with reasonable accuracy regardless of the total velocity achieved by the vehicle. Rockets tend to generate the additional velocity for longer ranges out of the earth's atmosphere. This is true even of low performance rockets such as the V-2. Indeed, rocket vehicles would suffer very severe aerodynamic heating penalties during acceleration if they did not generate most of their velocity outside of the atmosphere.

Since a general comparison of rocket performance is of interest here, and since the higher performance rockets of later chapters will be less sensitive to accurate drag allowances, drag will not be discussed further. For accurate performance estimates, it must be accurately calculated. For approximate comparisons, an allowance of about 2000 feet per second, regardless of total performance, is sufficient.

Gravity Losses

It is possible for rockets as well as people to expend energy without doing useful work. If a rocket is sitting vertically on a test stand with thrust exactly equal to its weight, if released, it will neither rise in altitude (increase potential energy) nor pick up velocity (increase kinetic energy). It does no useful work because its thrust has been countered by the force of gravity, although it may expend great energy supporting itself, a job done by the test stand before the engine was ignited. While a rocket in flight is burning, part of its thrust will be nullified by the component of gravity which lies along the thrust axis, and will not be available for acceleration. This effect is known as gravity loss.

The gravity loss can be seen by Figure 2-1 to be a function of the flight path angle. It may be expressed in terms of the additional velocity the rocket would have achieved if gravity had not existed as:

$$\Delta V_g = -g t_b \sin \gamma \qquad (2\text{-}5)$$

where ΔV_g = velocity loss due to gravity in feet per second; t_b = rocket burning time in seconds; and γ = flight path angle

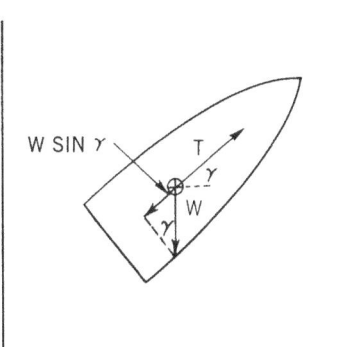

Fig. 2-1. Forces during motor burning.

measured from horizontal. In practice, the flight path angle varies throughout burning, and a weighted average value must be used in Equation 2-5. For a burning time of 60 seconds and average flight path angle of 60 degrees (approximately V-2 values), Equation 2-5 gives 1300 feet per second as the velocity loss due to gravity. As in the case of atmospheric drag, an allowance of a few thousand feet per second will cover the gravity loss during motor burning for most rocket vehicles. It is clear that the extremely short propulsion time of guns results in no appreciable gravity loss.

Airplane Lift/Drag Ratio

Most conventional forms of transportation spend only a small part of their effort in stopping and starting, but must use large amounts of energy during mid-course. Airplanes fight gravity incessantly. In level, cruising flight at constant velocity, they do not increase either kinetic or potential energy, hence, always experience a gravity loss. During this process, however, they do perform useful work since a force—the drag—is moved through a distance. Therefore, the energy required to produce the work done by an airplane while cruising is given by:

$$E_A = Ds \qquad (2\text{-}6)$$

where E_A = airplane energy required in foot-pounds; D = drag in pounds; and s = distance (range) in feet.

Although the atmosphere represents a penalty to guns and rockets, it is put to useful work by the airplane. Airplanes generate more lift than drag, and a measure of airplane cruising efficiency is the lift/drag ratio. Since lift is equal to weight in level flight:

$$E_A = \frac{ws}{L/D} \qquad (2\text{-}7)$$

where L/D = ratio of lift to drag. This energy may be expressed as a fictitious velocity at which kinetic energy would be the same as Equation 2-7. Hence, with the aid of Equation 1-4:

$$\Delta V_{EN}^2 = \frac{2g_o s}{L/D} = \frac{2g_o t_b V}{L/D} \qquad (2\text{-}8)$$

where ΔV_{EN} = velocity equivalent of energy used in feet per

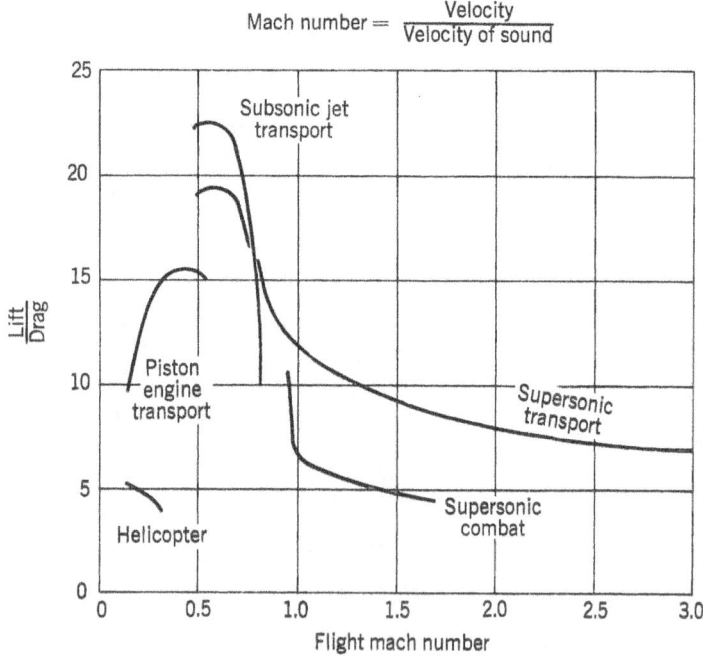

Fig. 2-2. Airplane lift/drag ratio.

second; t_b = time of flight in seconds; and V = cruising velocity in feet per second. Equation 2-8 may be used to compare
energy requirements of aircraft and rockets since ΔV_{EN} may
be compared to rocket velocity requirements previously given.

Airplane L/D values range from 25 for a highly efficient
subsonic aircraft to 14 for a typical 1965 jet transport to eight
for a supersonic transport. A curve of typical current values is
shown in Figure 2-2. For a modern jet transport with L/D of
14, Equation 2-8 gives a velocity of 8540 feet per second for a
range of 3000 miles. Since 5000 feet per second is roughly the
maximum velocity of the V-2, it is clear that mid-course energy
penalties of airplanes are similar to rocket velocity requirements.

Surface Transportation "Lift/Drag" Ratio

Although it is obvious airplanes must always fight gravity,
it is not as obvious that this is also true of all forms of surface
transportation. Although the ground or water supply the lift
force directly, the vehicle must be moved against the friction
caused by this force. In addition, atmospheric drag is always
present in terrestrial surface transportation.

The ratio of the weight of vehicle to the force required to
move it is the ground and sea transportation equivalent of aircraft L/D. Effective L/D's of modern automobiles are shown
in Figure 2-3. Air drag and large amounts of internal friction
combine to make the "L/D" of an automobile not much
different from that of an airplane.

The effective L/D of a ship is determined mostly by the
wave-making resistance of the water. This depends both on
the velocity and the length of the ship. Figure 2-4 shows typical values for different classes of ships. They range from over
300 for cargo ships and tankers, to 60 for large fast ships such
as cruisers, to 22 for destroyers at high velocities. Ships are
not greatly different from automobiles or airplanes, although
the cargo ship, apparently one of the most efficient ways to
operate in the presence of a gravity field, attains high "L/D"
but at the price of long travel time.

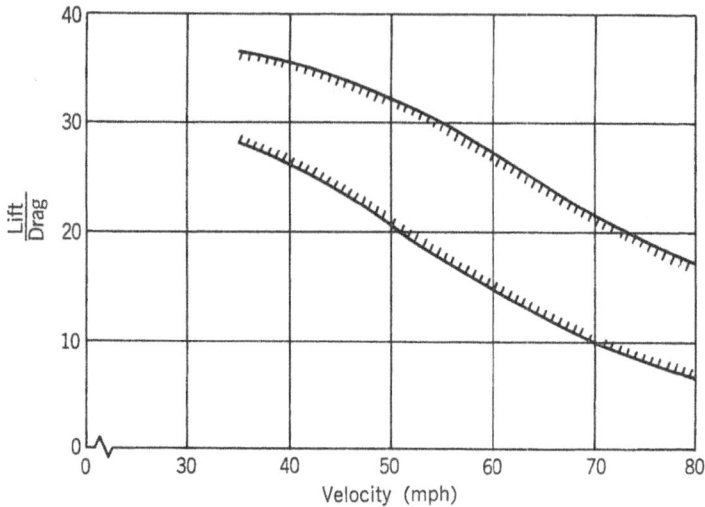

Fig. 2-3. Automobile lift/drag ratio.

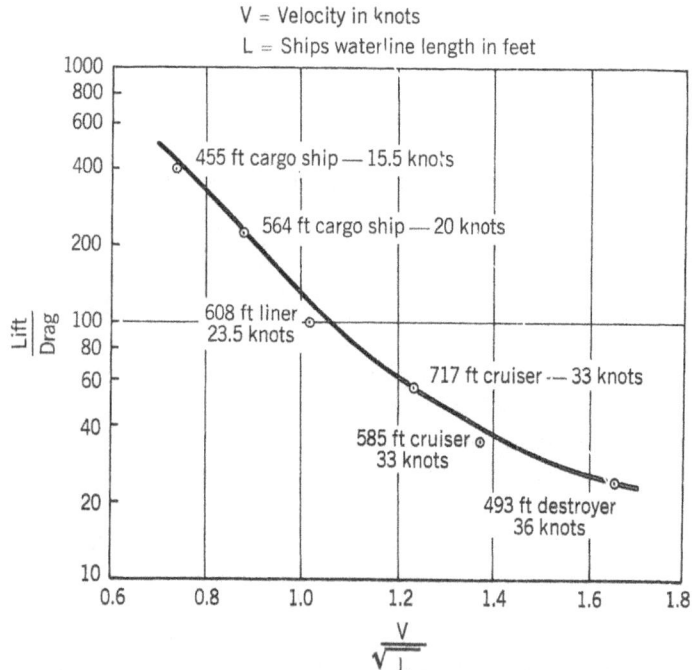

Fig. 2-4. Ship lift/drag ratio.

Equation 2-8 applies equally well to the cruising mode of all forms of transportation. It gives 2020 feet per second for an automobile with "L/D" of 25 traveling 300 miles, and 2600 feet per second for a cargo ship with "L/D" of 300 traveling 6000 miles. The energy requirements of normal terrestrial transportation are equivalent to rocket velocities of several thousand feet per second.

Regions of Travel

The various regions of gun and rocket travel can now be described as follows. The gun creates mostly kinetic energy and experiences negligible drag or gravity losses during its starting period. The rocket creates both potential and kinetic energy during the starting period, but experiences appreciable drag and gravity losses in the process. The gun projectile experiences severe drag losses during mid-course while the very long range rocket experiences none. Long range guns utilize trajectories with high enough peak altitudes to get some relief from drag, while very short range rockets have losses similar to guns. Neither device uses propulsion during mid-course and neither attempts to slow down when stopping. The remaining kinetic energy is delivered to the target along with the rest of the warhead.

The various regions of terrestrial transportation, then, differ greatly from guns and rockets. A relatively small amount of energy is devoted to creating potential or kinetic energy on starting, although in high velocity airplanes, the amount is not negligible. The mid-course portion consists of a large, continual gravity loss, compounded by atmospheric and/or water drag for land, sea, and air transport, but aided by atmospheric lift in the case of airplanes. Stopping energies are small and easily dissipated. Some initial kinetic energy is recovered by coasting to a stop in all cases. The airplane usefully recovers its starting potential energy on descent at the end.

The comparison of rockets with other forms of propulsion will require reference to these various different regions of transportation.

Solid Propellant Rockets

Most of the history of rocketry is the history of powder rockets. They have been known for centuries, and their origins are lost in Chinese antiquity. A reference to "arrows of flying fire" in A.D. 1232 is the earliest known presumed description of rocket usage. It is not known how they were invented, but any modern engineer can easily imagine the process. With gunpowder as tricky as it was and with virtually no technical understanding, no doubt a great deal of surprise was involved in producing the first rocket. The real problem, more psychological than technical, must have been in doing it the second time.

We will not attempt to trace the early history of rocketry. It is one of the great stories of human curiosity and ingenuity; neither more nor less stirring than many other developments during those centuries. For hundreds of years, rockets and guns were highly competitive. Particularly during the Nineteenth Century, rockets became prime bombardment weapons. They were substantially developed by the British General Sir William Congreve who became interested after Indian troops successfully used rockets against the British prior to 1800. They were used in the burning of Copenhagen in 1807 and found their way into the American National Anthem after being used in the bombardment of Fort McHenry.

The various early powder rockets were simply gun powder mixtures tamped into cases, frequently made only of paper, with a tapered hole down the center to permit the gases to escape. Various means, including long trailing sticks, were used to stabilize the flight. The range achievable by rockets and guns was not greatly different. The accuracy, however, introduced into guns during the latter part of the Nineteenth Century by the use of spinning projectiles from rifled barrels, was much better than that achievable with rockets until modern technology introduced aerodynamic dispersion reduction techniques and guidance systems. For about half a century terminating with World War II, rockets were virtually unused and the gun reigned unchallenged.

The heat release of the black powder commonly used in the Nineteenth Century was about 540 gram-calories per gram. Equation 1-12 would give an exhaust velocity of 7000 feet per second (I_{sp} = 218 seconds) for this if burned with perfect efficiency. When Goddard first measured the performance of rockets in use in 1914, he obtained exhaust velocities of only about 1000 feet per second, or an efficiency of only about two per cent. Furthermore, the rockets contained only 25 per cent fuel weight. A single-stage rocket without payload could only achieve a velocity of about 280 feet per second (see Figure 1-6), and consequently a maximum range according to Equation 2-2, of only 2440 feet (less than one-half mile). In previous centuries, no one had derived the rocket equation, so the manner of improving the situation was not even suspected.

During the Twentieth Century, the rocket equation was derived independently and published by Tsiolkovskiy in Russia in 1903, Goddard in the United States in 1919, and Oberth in Germany in 1923. Increases in performance by cutting down rocket weight as well as by increasing the energy of the reaction became of interest.

Goddard's Early Solid Rocket Experiments

Goddard derived the basic equation in 1912-13. He then set about to measure the efficiency of the current rockets with the results mentioned above. He reasoned that better energy conversion was possible, and that the way to do this was to burn the powder at high pressures and to expand it through a DeLaval nozzle to increase its velocity. The DeLaval nozzle was invented by a Swedish engineer, Gustav DeLaval, just for the purpose of converting thermal energy to kinetic energy. It had been used in DeLaval's steam turbines since about 1888. While he was at it, Goddard switched from black powder to smokeless powder, which had about twice the energy release.

Goddard achieved better than 50 per cent efficiency with the above changes. He measured exhaust velocities of over 7500 feet per second (I_{sp} = 233 seconds). The nozzle curves of Figures 1-2 and 1-4 show this clearly, although they do not extend low enough in pressure to indicate the very low efficiency of the earlier rockets. The theory behind Figure 1-4 was

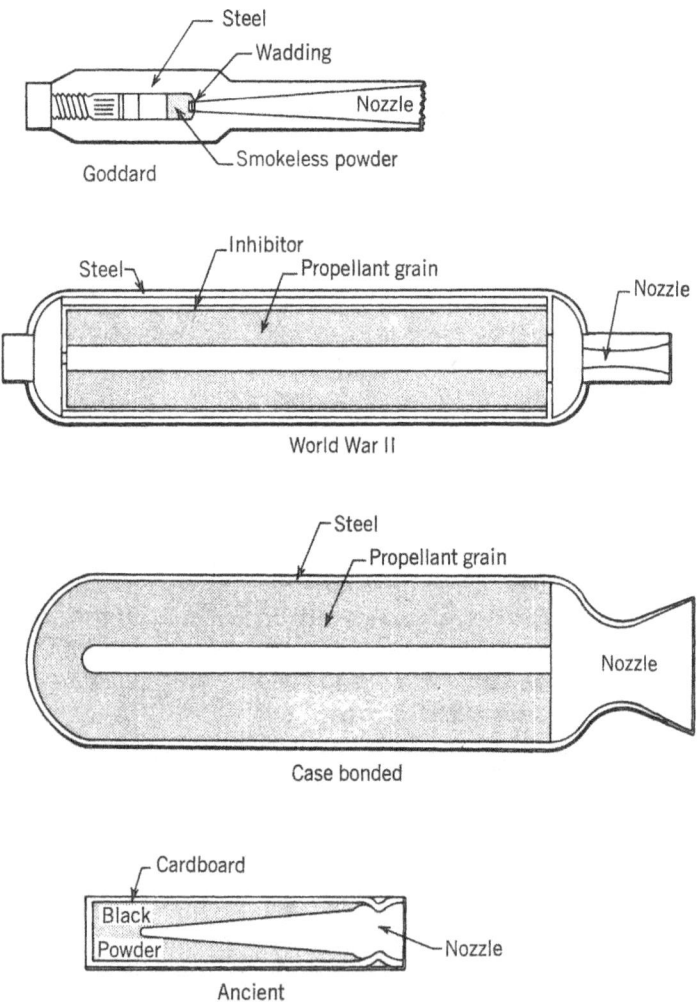

Fig. 2-5. Solid-propellant rockets.

not known to Goddard, and when he ran tests in a vacuum, he was surprised to find the thrust even higher. He incorrectly attributed this to ignition differences.

The use of modern explosive chemistry in rockets developed steadily after Goddard's experiments. Figure 2-5 shows a comparison of an early powder rocket, a Goddard rocket, a type of solid grain used during World War II, and a modern case-bonded solid propellant rocket.

It can be seen from Figure 1-4 that the chamber pressure should be at least 20 times atmospheric pressure to obtain energy efficiencies of over 50 per cent. Since sea level pressure is 14.7 pounds per square inch, a chamber pressure of 300 psi is sufficient to obtain reasonable energy conversion.

Black powder burns rapidly (explodes) at all pressures. One of the advantages of smokeless powder is that it burns slowly at atmospheric pressure, and only burns rapidly at the pressures developed in gun chambers. These pressures range from 20,000 to 50,000 psi, or 1300 to 3500 atmospheres. Thus, smokeless powder as used in guns burns at several hundred times the pressure necessary for rockets. Goddard's experiments burned gunpowder as it is used in guns (see Figure 2-5), and the thick steel cases were more typical of guns than of flying devices.

It is possible to formulate solid propellant charges with a wide variety of burning rates and burning pressures. The typical World War II rocket shown burned at 2000 psi pressure, and required extra weight of insulation inside the case for protection from the combustion gases. The modern case-bonded grain not only burns at pressures of 300–500 psi, but the bonding of case to grain means the grain itself protects the case from heat loads throughout most of the burning period. Thus, case weight in modern solid propellants has been reduced drastically from past values.

Other improvements through propellant chemistry have been important. It is necessary to prevent cracking in solid propellant grains since the extra burning surface exposed will cause extra combustion and an increase in pressure—sometimes catastrophically. This has always been a difficult problem over a wide temperature range. Modern grains, based on rubber partly for this reason, are a vast improvement, although further gains are highly desirable.

Post World War II Solid Rockets

By the end of World War II, solid propellant artillery rockets had re-invaded the province of some guns. Exhaust specific impulses of about 200 seconds were readily available. It was not uncommon to find 40 per cent of the rocket initial weight carried as fuel. Thus, a velocity increment of over 3000 feet per second was available.

Comparison of guns and rockets as bombardment weapons during the Nineteenth Century always involved a trade-off between the greater ammunition weight of the rocket and the greater launcher weight of the gun. Now, however, modern rockets are about as efficient as guns, as can be seen from Figures 1-2 and 1-7, realizing that guns are about 20 per cent efficient. Hence, rocket ammunition is at least as light as gun ammunition.

Advent of the airplanes as a major weapons system in World War II further influenced this comparison. It was possible to obtain far heavier fire power for the weight which airplanes could carry by means of rockets rather than guns. Today, rockets are the primary armament on most airplanes.

Since 1945, solid propellant rockets have been widely used in other applications. As first stage boosters for anti-aircraft missiles such as Nike, they provide a quick initial velocity increment. This is done with an acceleration of about 25-50 g's. Missile equipment can stand this but the thousands of g's inherent in guns would be unsuitable.

Under the simplifying assumptions of uniformly accelerated motion and zero initial velocity, the acceleration experienced by either guns or rockets in generating velocity is given by the following equation:

$$a = \frac{\Delta V^2}{2g_o s} \tag{2-9}$$

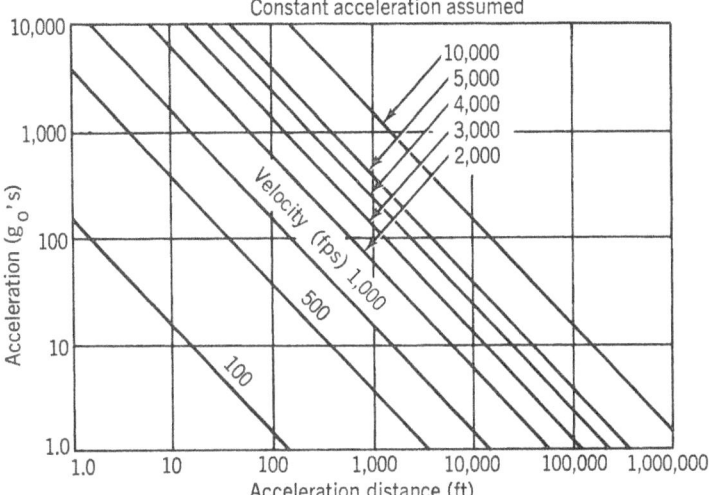

Fig. 2-6.   Required acceleration.

where a = acceleration in g_0's; and s = acceleration distance in feet. This equation is plotted in Figure 2-6. When several thousand feet per second are required, guns with barrels measured in tens of feet must place thousands of g's on their projectiles. Rockets avoid this by using thousands of feet to generate the velocity required.

The advent of nuclear warheads also changed the artillery rocket/gun comparison by generating a desire for relatively large payloads which could not stand the acceleration of guns. Consequently, the Honest John and Little John class of field artillery rocket came into being. Further understanding of the sources of rocket dispersion combined with modern aerodynamic knowledge and production control techniques have made these unguided rockets acceptably accurate. A similar application is the interceptor-launched Genie rocket with nuclear warhead.

Thus, after being completely overshadowed by guns for half a century, the powder rocket appears to be getting its revenge. Not only are its flashy liquid offspring plunging into space where guns cannot follow (unless carried by rockets), but its solid propellant derivatives are highly competitive in short range artillery usage.

Liquid Propellant Rockets

Liquid propellant rockets are a direct outgrowth of the rocket equation. Of the propellants known to rocket experimenters during the 1920's, liquid propellants had the highest energy release and hence those of low molecular weight had potentially the highest specific impulse. As soon as the importance of reducing rocket weight was recognized, attempts were made to decrease the empty weight of solid rockets. Since the propellant tank of a solid rocket is also the combustion chamber, it is pressurized when the rocket fires. One approach to saving weight is not to store all of the propellants in the combustion chamber. One can envision a small chamber into which solid propellant pellets are repeatedly fed and fired much like a machine gun. Attempts were made to build such rockets and Goddard obtained a number of patents on mechanisms. The end point of such ideas is to pump the propellant in continuously as a liquid.

Although the name of the inventor of powder rockets and the date of his triumph are lost in antiquity, we know that on March 16, 1926, Goddard launched the first liquid rocket. Its peak altitude was 41 feet and it landed 184 feet away. Four decades later, the Saturn V moon rocket will reach a height

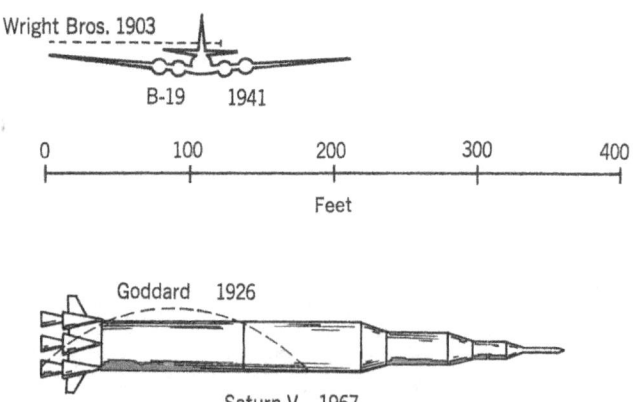

Fig. 2-7.   Four decades of development.

double that distance just sitting on the launching pad. The same thing happened in aircraft development. The entire first flight of the Wright Brothers on December 17, 1903, covered 120 feet and could have taken place on the 212-foot wing of the B-19 bomber which was flying 40 years later (Figure 2-7). These facts must prove something about the problems of predicting the degree of technical progress likely to occur over a period of four decades.

Storable Liquid Propellant Rockets

Liquid rocket development was pursued in various ways after Goddard's initial success. His original experiment involved the propellants liquid oxygen and gasoline. Liquid oxygen is what is known as a cryogenic propellant. Cryogenic means the propellant is a gas at normal temperatures and must be extremely cold to be a liquid. For instance, oxygen boils at 298° F below zero. The handling of such cold liquids obviously requires many special techniques such as refrigeration and the use of insulation to hold down evaporation losses. Because of these difficulties, much early liquid rocket development turned in the direction of "storable" propellants.

Storable liquid propellants usually do not have as high a performance as cryogenic propellants, but their performance was often adequate. By the end of World War II, storable liquid rockets as well as solid propellants had begun to be used in artillery class applications. The table on the following page lists some of the storable propellants and gives the specific impulse achievable. A typical example of a storable liquid system was the Nike anti-aircraft missile, which used nitric acid as oxidizer and gasoline as fuel. The missile was launched with a solid propellant booster.

Other applications of storable liquid rockets came into being, sometimes boosted by solid propellants, sometimes not. The Wac sounding rockets were followed by the Aerobee sounding rockets. Both used nitric acid and analine as propellants. The Aerobee used a solid booster. The Corporal rocket,

Theoretical Propellant Performance
Equilibrium Flow (Frozen Flow)

OXIDIZER	FUEL	VACUUM $\epsilon = 40$			SEA LEVEL
		Mixture Ratio	Specific Gravity	I_{sp}(sec)	I_{sp}(sec)
Double base (JPN*)			1.62	(294)	(250)
Ammonium Nitrate	18% binder & additives		1.51	(226)	(192)
Ammonium Perchlorate	20% binder & additives		1.72	(278)	(236)
Ammonium Perchlorate	12% binder, 20% aluminium		1.74	(314)	(266)
Red Fuming Nitric Acid	Analine	3.10	1.38	300 (290)	255
	Kerosene	4.80	1.35	315 (303)	268
	Hydrazine	1.47	1.28	332 (326)	283
Hydrogen Peroxide	Kerosene	7.35	1.30	321 (313)	273
	Hydrazine	2.09	1.26	337 (325)	287
Nitrogen Tetroxide	Hydrazine 50/50 UDMH,	1.40	1.22	342 (324)	292
	Hydrazine	2.08	1.21	340 (318)	289
Oxygen*	75% Alcohol	1.43	1.01	328 (314)	279
	Ammonia	1.40	0.98	346 (335)	294
	Kerosene	2.67	1.02	354 (324)	300
	Hydrazine	0.95	1.07	368 (343)	313
	UDMH	1.65	0.98	364 (347)	310

* Cryogenic

when it went into service as a medium sized surface-to-surface weapon, also used nitric acid and analine as propellants. In general, storable liquid rockets served very successfully in artillery type applications during the decade following 1945.

Storable Liquid Propellants versus Solid Propellants

One of the advantages of liquid rockets is that the propellants are not stored in the combustion chamber and, presumably, the propellant tanks can be lighter in construction compared to solid propellant rockets. The situation is not as simple as this statement indicates. It is necessary to get the propellants from the tanks into the combustion chamber. This can be done either by a pump or by pressurizing the tanks to force the fluid to flow into the chamber. If pumps are used, the penalty of the weight of the pumps must be considered.

Because of the pump weight penalty and desire for simplicity, small liquid rockets frequently use the tank pressurization method. This involves a weight penalty for the pressurization system, and can easily turn out to be more of a basic structural weight penalty than that suffered by a comparable solid propellant engine. Pressure in the liquid tanks must be higher than that in the combustion chamber to force the propellants to flow into the chamber. Thus, pressures throughout a liquid rocket system would be higher than a solid rocket for equal combustion chamber pressure.

Liquid propellants tend to be less dense than solid propellants, so that an equal weight of liquid propellant requires larger tanks. Since modern solid propellants frequently burn at just as low a combustion pressure as those used in liquid engines and since solid propellant nozzles are usually lighter than liquid propellant thrust chambers, liquid rockets are often heavier than comparable solid rockets.

In many recent applications, storable liquid rockets have been replaced by later versions of solid propellant rockets. Although early storable liquid propellants had a higher specific impulse than comparable solid propellants, modern solid propellants match storable liquids in specific impulse (see table on page 46), and seem to have achieved ease of field handling and reliability in more convincing fashion. Later versions of the anti-aircraft missiles in the Nike series—Nike-Hercules and Nike-Zeus—used solid propellants in all stages, while the first version, the Nike-Ajax, used the storable liquid with solid booster previously described. The Corporal has been replaced by the Sergeant and Pershing. Although storable liquid rockets are finding it hard to compete with solid propellants in the artillery field today, they are being used extensively in space applications.

Cryogenic Liquid Propellant Rockets

The development work on cryogenic liquid rockets prior to World War II proceeded in various places in small fashion much the same way work on other propellant systems occurred. The potentially high performance of cryogenic liquid propellants was widely recognized. This is not surprising. The heat released by their combustion had been known to chemists for many years.

It is hard to claim that the scientific community of the day was leader in these developments. In Germany, Fritz Lang, producer of the science fiction movie, "Frau im Monde," funded the development of a liquid oxygen-gasoline rocket in 1929 by a team headed by Oberth, with the objective of launching it in connection with the premiere of the film as a publicity stunt. The rocket, however, was never delivered. In the United States that same year, Goddard's work received a firm boost from a genuine American hero, Charles A. Lindbergh, who interested Daniel Guggenheim in aiding Goddard financially so that he could continue his work.

The largest Goddard rockets, built in 1940, were 22 feet long, weighed 736 pounds including 500 pounds of propellants, and were pump-fed, using liquid oxygen and gasoline. With a fuel/initial weight ratio of 68 per cent, these rockets would have achieved a velocity of 1.14 times exhaust velocity (see Figure 1-6) if they had worked. Liquid oxygen and gasoline should give about 15 per cent higher specific impulse than smokeless powder. Goddard's liquid oxygen-gasoline engines produced only 125–140 seconds specific impulse, however, compared to 233 seconds produced by his earlier solid experiments. Even so, the high fuel fraction would have yielded good performance. Due to various technical problems, the maximum altitude achieved was about 300 feet on August 9, 1940, although a Goddard pressure-fed rocket had achieved about 8500 feet altitude on March 26, 1937.

The greatest impetus of all to liquid rocket research, however, was given by another non-scientific group of men in 1919, the same year that Goddard published his first paper. They were the framers of the Treaty of Versailles. In what must be considered one of the classics of miscarriage of arms control thinking, these diplomats planted the seeds for the development of the long range artillery rocket, the German V-2.

The treaty severely limited long range artillery and aircraft. The German Army wanted a way around the treaty. The treaty did not cover rockets. In a superb piece of imaginative technical development, the German Army first subsidized the rocket experimenters, then organized its own experimental proving grounds to apply scientific and engineering techniques to rocket development. The first successful V-2 roared over the Baltic Sea on October 3, 1942, 16.5 years after the first liquid rocket flight. It had a range of 118 miles. This was greater than any gun ever fired including the famous World War I "Paris Gun."

The V-2 used liquid oxygen and alcohol as propellants. It weighed slightly over 28,000 pounds and was approximately 68.5 per cent propellant by weight. By today's standards, the V-2 is, of course, insignificant; however, at the time of its development it was a remarkable achievement. It was a much larger rocket than anyone had ever attempted before—either liquid or solid propellant. It was made operational during World War II under extremely difficult conditions, yet used a cryogenic propellant—liquid oxygen.

In an attempt to get what was high performance in its day, the V-2 was the pioneering effort in large tubro-pump fed rockets in order truly to cut down propellant tank weight. The

V-2 and Goddard's pump-fed rocket had almost identical propellant mass fractions. But the V-2 had 68,000 pounds vacuum thrust and achieved a specific impulse of 244 seconds in a vacuum. It generated a maximum velocity of about 5000 feet per second. The exhaust power of the V-2 was 485,000 horsepower (see Equation 1-11). The nozzle expansion ratio was 18.2. From Figure 1-2, it can be seen that the V-2 was about 75 per cent efficient in converting internal energy generated to useful exhaust energy, so that the internal power release of the V-2 engine was about 650,000 horsepower. These numbers are almost three times the total shaft horsepower of the most powerful ocean liner ever built. Thus, the useful power generated by rocket engines is far higher than most other propulsive devices.

It is frequently stated that a tremendous engineering achievement was required to package all of this power into such a small space before a rocket flight could become possible. Yet, if one examines a V-2 engine or even more modern and more highly developed engines, they do not seem too unusual. The pressures contained are reasonable. Cooling adequately keeps the chambers at a proper temperature, and the pumping systems, although well refined, are not unusual compared to the rotating machinery which exists in the modern jet airplane.

The basic point is that the V-2 engine did not package three times the horsepower of the Queen Elizabeth into a desk sized volume. It is true that the rate of energy production is that high, but in the ship propulsion system, all energy released must be converted within the machinery on board from chemical energy to the actual mechanical driving of the propeller shaft. Since the energy must be converted from one form to another, it must all be handled, so to speak.

The rocket engine, on the other hand, is a direct energy conversion device much like a gun. In the process of expanding out the nozzle, the chemical energy release is directly converted to useful thrust. The pumps which pump the propellant into the chamber never have to handle this huge energy release. In fact, the V-2 pumps were driven with only

460 horsepower, or 0.07 per cent of the total energy release. It is only necessary to protect the rocket chamber and nozzle from those small amounts of energy which happen to be transmitted into them. The basic process is one of permitting the energy released to escape from the system in a useful fashion, and only a few per cent of it is in any way handled.

The V-2 was not even the highest horsepower device of its time. The rate of energy release of guns is higher. The 2240 pound projectile of the 16-inch naval gun requires 43 milliseconds to achieve 2650 feet per second muzzle velocity (57 foot barrel), and about 50 million horsepower is released.

As shown in Figure 1-7, rocket energy is not very useful for propelling low velocity devices. The gross tonnage of the Queen Elizabeth, for instance, is 83,673. If she operated at that weight (never precisely true), and a V-2 were mounted on her deck and fired, by expending all propellants the rocket would produce $3/10,000$ of a g_0 for 65 seconds. It would be able to produce about 0.5 feet per second velocity of the ship, assuming no water resistance. If the ship were already at cruising velocity, the rocket would have only about three per cent of the thrust necessary to hold her there. This is not a very spectacular performance for the expenditure of three times the ship's horsepower. The large weight penalties of the ship propulsion system are in a great measure due to its necessity for completely handling all energy. The rocket's ability to control large amounts of energy by avoiding most of it is the secret of its small size and weight.

The V-2 represents a good place to close Chapter 2. As a technical development for its day, it cannot be praised too highly. It was clearly a complete failure as far as its basic objective was concerned. Someone else won the war. It was, however, a turning point in the history of rockets. It was the first time rockets achieved a really *decisive* edge over guns. Although aircraft still reigned as a means of delivery of heavy payloads, the V-2 clearly foreshadowed the beginning of the rocket-aircraft competition. Prior to the V-2, rockets had trouble competing with anything. Since the V-2, almost everything has had trouble competing with rockets.

3

Orbital and Global Rockets
(Velocity up to 36,700 Feet Per Second)

Ballistics

When rocket range exceeds a few hundred miles, it is necessary to allow for the curvature of the earth in calculating performance. The performance expressions cannot be derived as simply as in the previous chapter. Not only must the geometry of the curved earth be considered, but the fact that earth's gravity decreases as the square of the distance from the center of the earth must be included. Long range rockets, once their engines are shut down and they are beyond any influence of the earth's atmosphere, behave the same way as any other celestial body.

In the Seventeenth Century, Johannes Kepler, working with observations made by Tycho Brahe without the aid of a telescope, deduced three laws describing the motion of the planets about the sun. They were:

1. The orbit of each planet is an ellipse with the sun at one focus.

2. The line joining a planet and the sun sweeps out equal areas in equal intervals of time. (The planet moves faster when closer to the sun).

3. The square of a planet's time of revolution about the sun is proportional to the cube of its mean distance from the sun.

A long range rocket follows the same type of elliptical path except that a portion of the path goes beneath the earth's surface and cannot be followed by the rocket. Figure 3-1 shows typical paths and the nomenclature to be used in describing them.

While Kepler's laws were based on Brahe's observations,

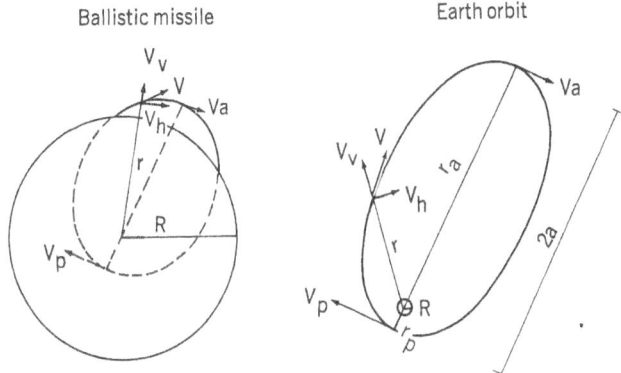

Fig. 3-1.   Elliptical orbit nomenclature.

Newton formulated an expansion of these laws utilizing his own laws of gravitation and motion. Newton concluded that Kepler's laws apply to any object moving under the influence of a single gravity field, but that the path may be any conic section, not only an ellipse.

Conic sections are curves formed by passing a plane through a cone. They consist of ellipses, parabolas, and hyperbolas. A circle is a special ellipse. The parabola is between the ellipse and hyperbola, and is the dividing line between trajectories which form closed orbits about the source of gravity (ellipses) and those which will never return (hyperbolas). The velocity regions of this chapter will cover only elliptical and parabolic earth trajectories.

Circular Orbits

The variation of gravity as a function of distance from Earth may be expressed as:

$$g = g_o \left(\frac{R}{r} \right)^2 \tag{3-1}$$

where g_o = standard acceleration of gravity (32.174 feet per second^2); r = radius from center of gravity in feet; and R = standard radius of the earth corresponding to the standard gravity (20.86 million feet).

According to Newton's First Law of Motion, a moving body will continue in a straight line unless subject to an external accelerating force. Centrifugal acceleration (given by V^2/r) is that required to hold a body on a circular path. The velocity of a satellite in a perfectly circular orbit about any body may be calculated, therefore, by equating this centrifugal acceleration with the acceleration of gravity at the satellite altitude. The result is:

$$V_c = \sqrt{gr} = \sqrt{\frac{g_0 R^2}{r}} \qquad (3\text{-}2)$$

where V_c = satellite (circular) velocity in feet per second.

If the standard values of gravity and earth radius are put in Equation 3-2, orbital velocity at the earth's surface is calculated to be 26,000 feet per second.

Because of atmospheric drag, obviously one cannot have a satellite at the earth's surface—not to mention that the earth is not perfectly round and does contain such hazards as mountains. If one imagined a perfectly smooth, spherical earth with no atmosphere, however, the minimum energy way to achieve infinite range would be to fire the rocket horizontally to orbital velocity. Thus, with a spherical earth, infinite range is obtained with horizontal velocity. In the short-range flat-earth approximation of the previous chapter, maximum range was obtained at 45 degrees elevation angle.

From the velocity expression of Equation 3-2, the time of revolution for one orbit of a satellite (its period) is seen as:

$$P_c = \frac{2\pi r}{V_c} = 2\pi \sqrt{\frac{r}{g}} \qquad (3\text{-}3)$$

where P_c = time to complete circular orbit in seconds. By using Equation 3-1:

$$P_c = \frac{2\pi}{\sqrt{g_0 R^2}} r^{3/2} \qquad (3\text{-}4)$$

Thus Kepler's Third Law is easily derived for circular orbits. The standard values of gravity and earth's radius yield 84.5 minutes as the period of a hypothetical satellite at the earth's surface.

Potential Energy

Since gravity is not constant, the potential energy expression previously given (Equation 2-3) must be modified. It becomes:

$$PE = wR \left(1 - \frac{R}{r} \right) = wR \left(\frac{h}{R+h} \right) \qquad (3\text{-}5)$$

If the altitude (h) is very small with respect to the earth's radius (R), this expression becomes the same as Equation 2-3. Although the gravity field extends throughout all of space as Equation 3-1 shows, the field becomes so weak at great distances that the potential energy has a finite maximum value given by:

$$PE_{max} = wR, \qquad r \rightarrow \infty \qquad (3\text{-}6)$$

If this were not so, it would mean that a gravity field generated by a finite mass had the capability to generate infinite energy—an obviously incorrect situation. Equation 3-6 gives 20.86 million foot-pounds per pound as the maximum gravitational energy of the earth's field, which is almost 12 times the gunpowder energy release of 1.75 million foot-pounds per pound given in Chapter 1. The fact that gravitational energy per pound is high compared to chemical energy release per pound was the source of several prognostications during the early Twentieth Century to the effect that space flight was impossible.

Escape Velocity

If a kinetic energy input is made which is just equal to the maximum potential energy, then the projectile will reach an infinite distance. It is said to "escape" from the potential field since it will follow a parabolic trajectory and not return to the gravity source, even though it is always under gravity influence. The velocity required for escape is given by:

$$V_E = \sqrt{2gr} = \sqrt{\frac{2g_0 R^2}{r}} \qquad (3\text{-}7)$$

where V_E = escape velocity from r in feet per second; and g = acceleration of gravity at r in feet per second^2.

Escape velocity, often called parabolic velocity, is always exactly $\sqrt{2}$ times circular orbital velocity at the same radius. If the standard values of gravity and earth's radius are put in Equation 3-7, escape velocity from the earth's surface is calculated to be 36,700 feet per second, neglecting atmospheric drag.

The Vis-Viva Law

In celestial orbits, potential plus kinetic energy remains constant as in the flat-earth trajectories of Chapter 2. From Equations 1-4 and 3-5 then:

$$KE + PE = \frac{w}{2} \frac{V^2}{g_0} + wR \left(1 - \frac{R}{r} \right) \qquad (3-8)$$

Since velocity and radius are the only variables in Equation 3-8, it is evident that the constraint placed on trajectories by the conservation of energy leads to velocity depending only on radius. One could hope to find simple expressions relating these terms.

In addition to the conservation of energy, angular momentum must be conserved just as the linear momentum discussed in Chapter 1. Angular momentum is the momentum which causes rotation of the trajectory. The velocity component perpendicular to the line connecting the body and the earth's center (see Figure 3-1) causes rotation. This is the "horizontal" velocity of Equation 2-1.

The conservation of angular momentum requires that a body at a great distance must have a lower trajectory rotation rate than if it were close. Kepler's Second Law, that equal areas are swept out in equal times, follows directly from the conservation of angular momentum. A more familiar example of angular momentum conservation is a twirling ice skater. The rate of spin will be faster with arms held close to the skater's body than when extended.

The point of an orbit which is farthest from earth is called the apogee. The point closest is called the perigee. (The prefixes apo- and peri- are similarly used for all celestial orbits, sometimes with Greek, sometimes Roman, suffixes. Hence, for Mars—apomartian, perimartian; for Sun—aphelion, perihelion, etc. General terms are apoapsis and periapsis. Common usage is not yet established in all cases.) The total velocity at both apogee and perigee is directed "horizontally," otherwise they would not be maximum and minimum distances (see Figure 3-1). Hence, the conservation of angular momentum yields:

$$V_a r_a = V_p r_p \qquad (3\text{-}9)$$

where the subscripts a and p refer to apogee and perigee respectively. This relation can be used with Equation 3-8 to relate perigee velocity increment to apogee:

$$\frac{V_p^2}{V_{Ep}^2} = \frac{r_a}{r_a + r_p} = \frac{r_a}{2a} \qquad (3\text{-}10)$$

where V_{Ep} = escape velocity at perigee; and a = semi-major axis of orbit. The subscripts a and p in Equation 3-10 can be interchanged to give an identical relation between apogee velocity and perigee. Equation 3-10 can be combined with Equation 3-8 to give the following expression:

$$V^2 = g_o R^2 \left(\frac{2}{r} - \frac{1}{a} \right) \qquad (3\text{-}11)$$

Equation 3-11 is a general expression relating the velocitiy and distance from the center of gravitation for all points on an orbit. It is known as the Vis-Viva Law, and much can be learned from it.

The condition for a circular orbit is r = a at all points. Equation 3-11 becomes Equation 3-2 for this case. Similarly, escape occurs when a → ∞. Then Equation 3-11 becomes Equation 3-7. The Vis-Viva Law shows clearly that for a given semi-major axis of orbit, the velocity is a function only of the radius.

The purpose of illustrating one derivation of the Vis-Viva Law was to show that the kinetic and potential energy relations result in simple, useful expressions. Furthermore, easy extensions from one gravitational field to another can be made. We may slightly re-write the Vis-Viva Law as:

$$\left(\frac{V}{V_{c_0}}\right)^2 = \left(\frac{2}{r/R} - \frac{1}{a/R}\right) \qquad (3\text{-}12)$$

Thus, if we know the circular velocity (V_{c_0}) at a standard radius (R) and calculate V/V_{c_0} and r/R ratios, they will be the same for all gravitational fields. Although all the derivations of this chapter are made as if for earth, the curves are plotted in ratio form and apply throughout the solar system, or beyond it for that matter.

The rate at which the radius vector sweeps out area at perigee, $r_p V_p/2$, applies throughout an orbit due to conservation of angular momentum. Since the area of an ellipse is $\pi a \sqrt{r_a r_p}$, Equations 3-7 and 3-10 may be applied at perigee to derive the orbital period for elliptical orbits, Kepler's Third Law. The result is:

$$P_{or} = \frac{2\pi a^{3/2}}{\sqrt{g_0 R^2}} \qquad (3\text{-}13)$$

where P_{or} = orbital period in seconds.

Minimum Energy Trajectories

In Chapter 2, simple flat-earth ballistic equations were presented. In this chapter, some useful expressions from celestial mechanics have been discussed. Much more complicated derivations are necessary for the velocity requirements of interest. These include filling in from short range to escape velocity, and involve interest in travel times and trajectory details as well. The Vis-Viva Law gives only the magnitude of velocity as a function of radius, but tells nothing of the direction of that velocity. The velocities required for orbit-changing maneuvers also depend on flight path angles. Total velocities will be presented by means of curves with the previous expressions used to aid in a feel for their nature.

Because high performance rockets are a recent development, and originally were thought to be an extremely difficult technical achievement, much interest has centered around the smallest size rocket possible for each mission. Hence, the minimum energy trajectories to reach a given range are of interest. As indicated in Chapters 2 and 3, the optimum angle

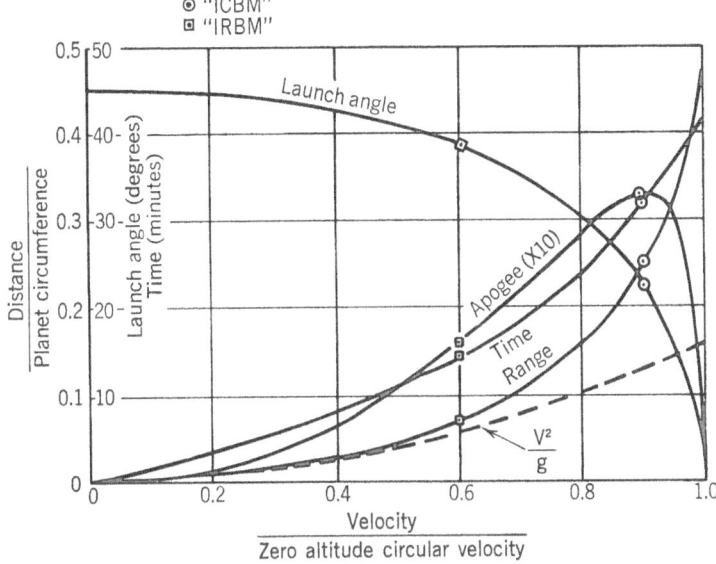

Fig. 3-2. Optimum ballistic missile trajectories.

for maximum range must vary from 45 degrees at short range to zero at long range. Figure 3-2 shows velocity required, angle of launch, maximum altitude, time of flight, and range for ballistic missiles fired on an idealized earth with no atmosphere. These curves are for optimum trajectories, i.e., at the angle shown the minimum velocity is required to achieve the desired range.

An ICBM has come to be defined as a ballistic missile with range approximately one-quarter of the way around the earth. Maximum altitude is reached when one travels precisely that distance and at that point the optimum angle of launch is exactly one-half 45 degrees. The velocity required, however, is 23,400 feet per second, about 90 per cent of satellite velocity. This is why intercontinental ballistic missiles and orbital launch vehicles are so closely related.

General Trajectories

In actual usage, one would not always use the minimum energy trajectories. It might be desired to launch at either higher or lower angles than optimum. This could be done in military applications to help confuse the defense or in space applications when special missions are desired, to reach extreme altitudes rather than range, or to use flat trajectories for special re-entry heating tests. Figure 3-3 gives velocity and launch angle as a function of range for non-optimum as well as optimum cases.

An ICBM of 23,400 feet per second is constrained to fly to its target by the shortest route and to re-enter at approximately 22.5 degrees entry angle. A rocket of orbital velocity capability, however, can reach any point on earth. Furthermore, it can come around either the short or the long way. Rockets with orbital velocity capability or greater have become known as global rockets. A large array of trajectories is available to global rockets, but they, too, have their limitations. Any rocket with velocity exceeding escape velocity will never return to earth. Hence, not only does Figure 3-3 termi-

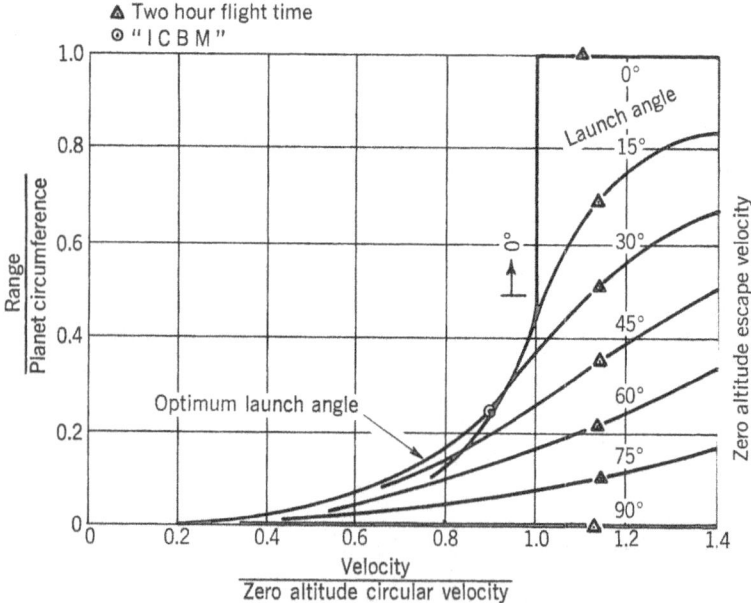

Fig. 3-3.   Global rocket velocities.

nate at escape velocity, but not all angles of approach are possible at all combinations of range and velocity. At 45 degrees launch angle, for instance, the rocket goes only half way around the earth, even at escape velocity.

The flight times become extremely long at the high velocities since the rocket's apogee becomes infinitely large at escape velocity. Flight times of two hours are spotted on Figure 3-3 to show these regions. This may be contrasted with the 32 minute standard ICBM flight time and the zero altitude satellite period of 84.5 minutes (see Equation 3-4). The latter means that the low orbit satellite form of global rocket can reach any point on earth in less than 43 minutes plus acceleration time penalties.

These various restrictions could be lifted by using some rocket velocity to change trajectories either at apogee or some other point subsequent to launch. This will not be discussed.

Hohmann Transfers

Orbital velocities have only been attained since 1957 and, as of 1965, the maximum velocities attained have only been slightly beyond earth escape velocity. It is reasonable, then, that we are extremely interested in minimum energy ways of achieving orbits, as well as ICBM ranges. It can be shown that the easiest way to achieve a given orbital altitude is by the procedure illustrated in Figure 3-4. The rocket is first put into an elongated ellipse called a transfer orbit. It has its apogee at the orbital altitude to be achieved. The perigee will be somewhere a few hundred miles down range from the launching point. Enough velocity beyond circular is put in at perigee to achieve the desired apogee and at apogee an additional velocity input is made to inject the payload into the circular orbit. This is called a Hohmann Transfer, after Walter Hohmann who was not an astronomer, as one might expect, but the city engineer of Essen, Germany. He showed that this type of transfer resulted in minimum velocity in a paper published in 1925.

In actual practice, the manner of achieving orbital altitude just described is not always used. The difference between a Hohmann Transfer and the use of an orbital transfer ellipse which covers shorter distance is usually not great. The shorter transfer is frequently more convenient from a guidance point of view, and it is not uncommon to make use of the less efficient trajectories.

The velocity requirement to achieve orbit may be calculated from the previous expressions by using Equation 3-10 for perigee velocity, Equation 3-9 to get from perigee to apogee velocity, and Equation 3-2 for orbital velocity. The amount added at apogee is obviously the difference between orbital and apogee velocities.

A plot of total velocity required to achieve circular orbits versus orbital altitude above the planet's surface is shown in Figure 3-5. One interesting aspect shown is that it is more difficult to achieve circular orbits at some altitudes than it is to escape. Achieving the altitude at which the orbital rotation of a satellite would be exactly synchronized with the earth's rotation so that the satellite would appear to hover motion-lessly over the earth, the so-called synchronous satellite, re-quires about six per cent more than escape velocity.

At first thought, the point that it requires more velocity to place a satellite in some orbits than to escape does not seem

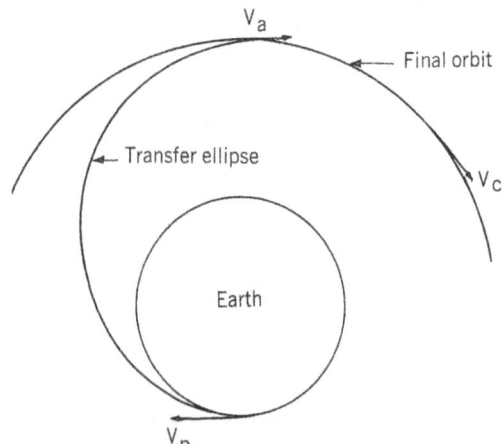

Fig. 3-4. Hohmann trans-fer.

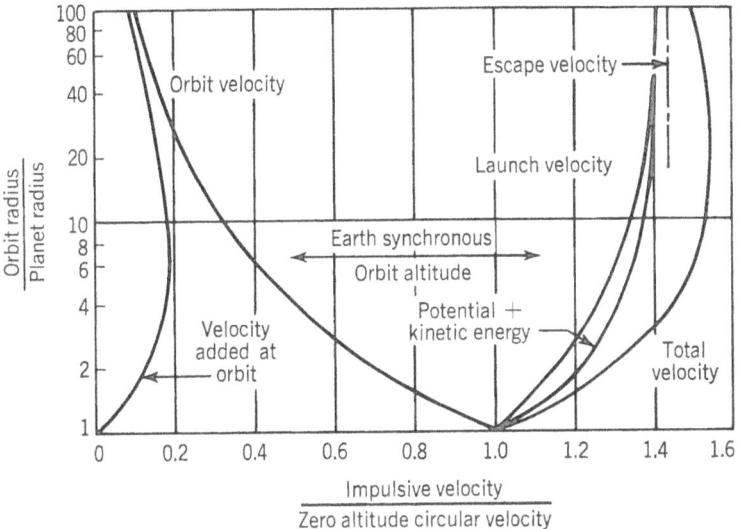

Fig. 3-5.   Velocities required to establish orbit.

plausible. The total potential plus kinetic energy per unit weight of the satellite in orbit, also shown in Figure 3-5, is always less than escape energy, and is given by (Equations 3-2 and 3-8):

$$PE + KE = R \left(1 - \frac{R}{2r} \right) \qquad (3\text{-}14)$$

Only this much energy is expended in placing the satellite in orbit. The final velocity input at apogee occurs, however, after the initial (perigee) velocity has been greatly reduced by the conversion of kinetic to potential energy. This is an illustration of a case where energy and velocity requirements differ. The energy utilized is less than escape energy. The velocity required to generate the energy, however, is greater than the velocity required to generate escape energy.

Other Planets and Satellites

The velocity region of this chapter includes up to Earth escape velocity, 36,700 feet per second. Hence, it does not include rockets capable of going to any other celestial body, except for hard impacts on the Moon. It does include the velocity region for rockets operating on bodies with escape velocity less than that of Earth. This includes four other planets, and all the natural satellites of the solar system. Everything in the solar system, in fact, but the four major planets, is included. Escape velocities and radii of bodies with less gravitational energy than Earth are as follows:

PLANET		ESCAPE VELOCITY (feet per second)	RADIUS (Earth radius = 1.0)
Earth		36,700	1.00
Venus		33,600	0.97
Pluto		32,700	1.1
Mars		16,400	0.53
Mercury		13,700	0.38
SATELLITE	(PLANET)		
Triton	(Neptune)	10,400	0.31
Ganymede	(Jupiter)	9,430	0.39
Titan	(Saturn)	8,900	0.39
Io	(Jupiter)	8,250	0.26
Moon	(Earth)	7,800	0.272
Callisto	(Jupiter)	7,450	0.37
Europa	(Jupiter)	6,900	0.23

24 Smaller Satellites

Thousands of Asteroids

Figures 3-2, 3-3, and 3-5 apply directly to all of these bodies. Except for Venus, which is close to a twin of Earth, and Pluto, which is so far away that its size is in considerable doubt, the largest velocity in the above table is that of Mars, and it is less than half the value of Earth. An Earth IRBM would be a global rocket on Mars. A V-2 would be a global rocket on the Moon.

Relatively low velocity rockets are thus capable of providing local transportation on almost all of the bodies of the solar system, even if much higher velocities should be used for interplanetary travel.

A case can be made that the human race was raised on the wrong planet. The Martian orbital velocity of 11,600 feet per second is only 10 per cent higher than the best exhaust velocity in the table on page 46. Normal chemical rockets would make excellent orbital transports and good escape vehicles on Mars. A Martian would probably view Earth, spacewise, much as we view Jupiter—a planet with a dense and forbidding atmosphere and too much escape velocity surrounded by an interesting satellite (or satellites) which can be explored for a reasonable energy expenditure.

Gravity Losses

The gravity losses during engine burning of high velocity vehicles are usually no greater than low velocity vehicles since actual trajectories, once out of the atmosphere, tend to be programmed closer to the horizontal as the velocity increases. Hence the loss due to a component of gravity along the thrust axis decreases. Furthermore, if the horizontal velocity is an appreciable amount of circular velocity, the gravity loss term must also include the effect of centrifugal force. A satellite is obviously not feeling a gravity loss. The effective gravity is:

$$g_{ef} = g - \frac{V_h^2}{r} = \frac{V_c^2 - V_h^2}{r} \qquad (3\text{-}15)$$

The first expression reduces to Equation 3-2 for $g_{ef} = 0$. The g in Equation 2-5 should be replaced by g_{ef} from 3-15 in the general case.

When the horizontal velocity exceeds circular, effective gravity becomes negative, i.e., the projectile's apogee will exceed its current radius. In this case, if the thrust were pointed toward the earth, there would be a "gravity" penalty.

In the general case, gravity (or more accurately gravity plus centrifugal force) losses during burning occur whenever the effective gravity has a component along the thrust axis. For most velocities beyond orbital velocity, the thrust is aligned along the flight path, since this is the most efficient way to increase kinetic energy. The flight path is usually close to horizontal, so the gravity loss is low. We can, therefore, get a good feel for propulsion requirements for long range missions by simply using the allowance (a few thousand feet per second) for gravity losses during burning discussed in Chapter 2.

Energy Comparisons

Some comparisons of energy required for various missions are instructive at this point. It was shown earlier in this chapter that the maximum gravitational potential of the earth was 20.86 million foot-pounds per pound. This, of course, is exactly the kinetic energy per pound of an object travelling at escape velocity, 36,700 feet per second. (The velocity at which kinetic energy per pound is equal to gunpowder is about 10,600 feet per second). Orbital energies per pound are one-half the escape energy value.

Although energies six or twelve times gunpowder energy are impressive, the price of such energy is not. Liquid oxygen and kerosene are chemical propellants of the gunpowder class of energy release, and the combination costs about 3 cents per pound. Hence, the price of energy equivalent to orbital energy is about 20 cents per pound. The electrical industry measures energy in kilowatt-hours and one million foot-pounds per pound is equal to .377 kilowatt-hours per pound. Thus, the orbital energy of one pound is about four kilowatt-hours. The commercial price of electrical energy in Washington, D. C., in 1965 was about one cent per kilowatt-hour. Hence, four cents is the price of modern electrical energy equivalent to one pound in orbit.

As shown in Chapter 2, a modern jet transport airplane with L/D of 14 uses the energy equivalent of 8540 feet per second in travelling 3000 miles (see Equation 2-8). In order to travel around the earth on a great circle course (24,800 miles), the airplane would utilize energy equivalent to 24,600 feet per second. Hence, a jet transport circling the earth would utilize about the same energy as required for an earth satellite.

Kerosene burned in air releases about 20,000 BTU per pound. This is 15.6 million foot-pounds per pound, or about nine times gunpowder energy release. The energy release is so high because about ten pounds of air is consumed per pound of kerosene. Kerosene and air combined are not much different from gunpowder on a total weight-consumed basis. If the airplane could convert the energy of its fuel and air burned to useful work with 100 per cent efficiency, a weight of fuel equal to the empty weight of the airplane would be more than adequate for circling the earth non-stop. Modern transports do carry fuel about equal to empty weight, but the overall energy utilization process is only about 25 per cent efficient, so the best of airplanes must be refueled several times for such a flight.

If one uses Equation 3-8 to calculate the velocity equivalent of an automobile of "L/D" = 25, travelling 3000 miles, the result is 6400 feet per second. When burned in air, kerosene, gasoline, and diesel oil all have energy releases within one per cent of each other. Hence, a pound of gasoline used with complete efficiency has about 25 times the energy release necessary to drive a pound of automobile across the United States, and five times as much if used with 20 per cent overall efficiency. Hence, one should be able to design an automobile to cross the United States non-stop by carrying about 20 per cent of its weight in fuel.

Most cars today are designed to carry about three per cent of weight as fuel, so that they must be refueled a minimum of seven times to cover this range. The situation is somewhat worse because each time a car decelerates from cruising velocity, its kinetic energy is dissipated by the brakes and must be replaced from the fuel when picking up velocity again. Hence, the actual fuel consumed is greater than these estimates indicate.

Because normal automobiles are designed with only three per cent fuel capacity, fuel supplies in convenient places are necessary. This is aided by oil fields and gasoline refineries located at various places around the country, but it is still necessary to use special tanker vehicles to deposit fuel at service stations ahead of time. An automobile crossing the United States represents a multiple refueling operation, with the payload carrier of only three per cent fuel weight. The other vehicles, tank trucks which have about 75 per cent fuel weight, are equally necessary.

The same sort of calculation applied to a cargo ship of "L/D" = 300 and 12,000 mile range yields the point that one pound of diesel oil contains 75 times the energy required to

move one pound for the trip, or 15 times the necessary energy
if utilized at 20 per cent overall efficiency. Hence, about seven
per cent fuel tanks are necessary. Ships are usually designed
with 5–10 per cent fuel weights. Cargo ships can get long
ranges, but high velocity combat ships, with L/D down
around 30, are much more limited in range at high velocity.
A high velocity luxury liner will use most of its fuel supply
covering only 3000 miles.

Many ships operate as multi-refueling devices, particularly
in naval operations. The same high fuel weight vehicles necessary for land travel also ply the oceans, either to refuel combat
ships at sea, or carry oil to supply points for later use. Interestingly enough, these ocean tankers carry about 75 per
cent of weight as fuel, quite comparable to similar land
tanker vehicles.

Further energy comparisons will be made throughout the
book. The point is simply that although the *velocities* of space
travel are 100 to 1000 times higher than terrestrial travel, the
energies consumed are comparable to long range aircraft, and
not greatly higher than the best surface transportation. Surface transports get away with small fuel loads only because
they use refueling operations extensively. *The problem in
space travel is not that the energy required is high. The problem is to learn to generate it efficiently, package it, and carry
it to space.*

Solid Propellant Rockets

It seemed intuitively obvious to many people that, although
solid propellant missiles had become very useful in anti-
aircraft and artillery rocket roles, clearly the long range ballistic missile would remain the province of the liquid propellant vehicle. Superior specific impulse was the basic reason
cited. As can be seen from the table on page 46, however,
solid propellant specific impulses are not much lower than the
liquid propellants which were under extensive development
prior to the launching of Sputnik. It was inevitable, then, that

attempts would be made to apply the reliability which solid propellants had achieved in artillery rocket applications to long range ballistic missiles.

Solid Propellant Boosters

It is usually possible to increase the acceleration available from a solid propellant without additional weight penalty. This is because the entire solid grain is pressurized, and more thrust can be obtained by changing nozzle exit size and propellant burning rate with no substantial change in total weight of the engine. With liquid propellant turbo-pump vehicles, more thrust can be obtained only by adding more thrust chamber, turbo-pump, etc., weight. These differences in engine characteristics mean that solid propellant ballistic missiles should be designed for greater take-off acceleration than comparable liquid ballistic missiles. This reduces the gravity loss penalty.

The ease of obtaining high thrust/weight ratio in solid propellants also makes them unusually adapted for supplementary booster applications when a large vehicle is given a modest additional boost with small auxiliary solid engines. The Thrust Augmented Thor, which uses solid engines as auxiliary propulsion at take-off, benefits from the extra acceleration in reducing gravity loss. This same characteristic results in the number of stages for optimum performance being larger with solid propellants than with the liquid propellants.

Solid Propellant Ballistic Missiles

The initial impetus for a large solid propellant ballistic missile came more strongly from the Polaris program than any other source. This was because shipboard use puts unusually stringent requirements on ease of handling and storability of propellants. For the submarine launched missile, there was a tendency to take a performance penalty in order to make use of the superior handling advantages of solid propellants. This performance penalty did not turn out to be very great, if it existed.

By the time of Polaris, continual developments in the chemistry of solid propellants had led to specific impulses as high as 240 seconds (see table on page 46). Modern case-bonding techniques and relatively low-pressure burning grains had led to very reasonable structural weights for solid propellant cases. Improvements in grain structural characteristics permitted large size (five-foot diameter) grains without cracking problems, and the thermal conditioning available on shipboard meant that grains that withstood wide temperature variations were unnecessary. A greater number of stages for a given mission was feasible due to the greater simplicity of solid propellant missiles. They were also considerably more compact for a given weight.

Since the solid propellants of the table on page 46 indicate exhaust velocities of 7500–8500 feet per second (I_{sp} of 240–260 seconds), an IRBM must be designed for an impulsive velocity of about 2.5 times exhaust velocity, and an ICBM about four times exhaust velocity. Figure 1-5 then gives about five per cent payload weight for the IRBM, and about one per cent for the ICBM, if properly staged. These are reasonable payload weights for weapons (recall that a bullet frequently weighs 0.1 per cent the weight of the gun).

The same factors which made solid propellant engines such heavy choices for artillery work have been applied to long range ballistic missiles. As a result, solid propellant engines also seem to have taken over the ballistic missile field, with Polaris submarine-launched missiles and Minuteman ICBM's as replacements for Atlas and Titan.

Minuteman and Polaris represent an interesting point in the development of rocket weapons. Although they are long-range strategic weapons which compete with airplanes as the means of strategic delivery, they actually both look and act more like very long range guns than flying aircraft. The artillery now has global range. It required rockets rather than guns to achieve this.

Future Large Solid Engines

The success of solid propellant engines of the Polaris and Minuteman size has led to development programs for even larger engines. Some are sketched in Figure 3-6. A motor of ten-foot diameter is under development as a first stage for the Titan III vehicle. In order to permit reasonable transportation of these motors, they are built in segments each of which is about ten feet long. Even so, each loaded segment weighs almost 40 tons. A five-segment unit, including nozzle, front closure, attachments, etc., will have a weight of about 240 tons.

Handling solid propellant charges in small units is not new. The powder charges in heavy artillery consist of several separately packaged units. In the 16-inch naval gun previously mentioned, six such units comprise a full charge. In the case of large solid motors, it is currently believed better to design the whole motor including outer case in segments, than to place multiple charges in a single case. Case bonding can then be properly handled in each segment, and good design of the structural joints between segments results in only a small weight penalty compared to a single case.

The largest solid motor currently supported is one of 260 inches (21 feet 8 inches) diameter. Even if built in 20 foot segments, each unit would weigh about 280 tons. The maximum clearance permissible by rail travel is about 13 feet. The practical weight limit is about 100 tons. Hence, even if segmented, 260-inch diameter motors will probably be moveable only by sea. Current plans are just that. The largest motor sketched in Figure 3-6 contains 1500 tons of propellant and produces 7 million pounds of thrust. It would be as heavy as a modern destroyer-escort. At a specific impulse of 260 seconds, its useful kinetic energy release would be about 53 million horsepower, and total internal energy release about 70 million horsepower. This rocket would be about the same horsepower as the 16-inch naval gun, but could produce this power about 3000 times longer. It is not certain yet that solid propellant engines of this size will be built, but active research

Diameter

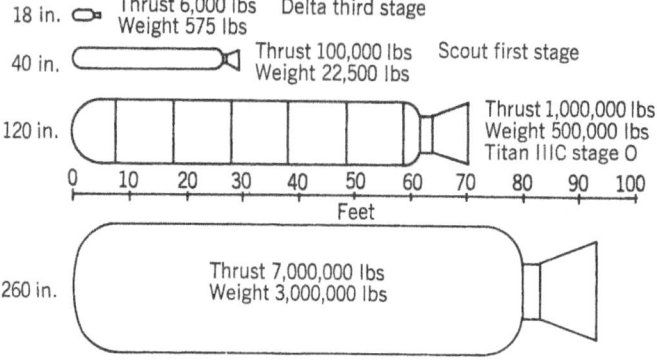

Fig. 3-6.   Large, solid-propellant motors.

and development leading to them is under way. There is little doubt as to their feasibility.

There will be little further discussion of solid propellants in this book. The search for higher specific impulse leads to both more energetic reactions and lower molecular weight propellants (see Equation 1-14). The low molecular weight elements appear as liquids or gases rather than solids, and hence it would seem that liquid propellants have a basic future advantage. If true, then the inexorable relations of the rocket equation will limit solid propellants to low velocity increments until some presently unforeseen formulation is produced. Solids were able to compete in performance with liquids in the ICBM velocity range, but it is quite a different matter when far higher velocity increments are necessary, and the competing systems are high-energy liquid or nuclear rockets.

Liquid Propellant Rockets

Development of liquid rockets continued in the United States and elsewhere following World War II. This was not surprising considering the spectacular entrance of the V-2 onto the weapon delivery scene. For awhile, artillery rocket applications were almost the only activity. Inevitably, storable and cryogenic liquid propellant rockets were applied to long range ballistic missile and space programs.

Storable Liquid Propellants

Cryogenic propellants were successfully used by the V-2. A really desirable military ballistic missile force, however, is capable of continuous alert. Although this can be done with cryogenics, it is, at best, an awkward situation. Consequently, the same requirements that led to intense interest in solid propellant ballistic missiles led to active consideration of storable liquid propellants in these applications.

The preivous storables, such as nitric acid and analine, had specific impulses of the order of 225 seconds. This does not make an attractive ICBM. Other propellant combinations (see table on page 46) are capable of matching the performance of liquid oxygen and kerosene. One of these combinations is hydrazine and nitrogen tetroxide. This propellant combination has now been highly developed and is used in the Titan II ICBM.

The list of storable, solid, and cryogenic propellants in the table on page 46 is a small indication of the difficulty of keeping abreast of rocket propellant possibilities. At any one point in time, it is easy to conclude that cryogenic performance exceeds storable, or that solids cannot exceed a certain value. But this decade's storable or solid may well exceed last decade's cryogenic in performance, and rocket engineeers are in a state of constant flux, attempting to understand the best applications at any given time.

Hybrid Propellants

It may be possible to obtain some of the advantages of both liquid and solid propellants by making a rocket engine using both. In this case, a liquid propellant is pumped through a chamber of solid propellant. Particularly if the propellants are hypergolic (the term applied to propellants which react immediately on contact with each other and therefore require no ignitor), such a rocket can be turned on and off at will. It may be simpler than two-liquid systems, and can use propellant combinations which pure liquid or solid rockets cannot.

Hybrid rockets may have large use in the future, but they will not be discussed further here. They are a different way of building chemical rockets, and their performance will not be markedly different from other chemical rockets. If they do combine all the advantages of liquid and solid propellants, their use will become widespread in those regions where chemical propellants are pertinent. If they turn out to combine all the disadvantages of liquid and solid propellants, their use will not be so widespread.

As of 1965, hybrid rockets have received only modest attention in the United States. France, however, has pursued them with substantial effort. The first rocket vehicle using a hybrid engine was launched by the French off the southern coast of France on April 25, 1964.

Cryogenic Liquid Propellant Rockets

Development of cryogenic rockets has been intense in the last two decades. By 1954, relatively large liquid rocket engines using liquid oxygen and kerosene were under intensive development in the United States as booster rockets for the Navaho long range ram-jet vehicle. One of these engines had approximately 150,000 pounds of thrust, or three times the thrust of a V-2 engine. When the United States decided to develop a ballistic missile force on an extremely rapid time scale, this engine became the principal workhorse of the program. The initial ballistic missile program consisted of two intercontinental ballistic missiles—the Atlas and the Titan, and two intermediate range ballistic missiles—the Thor and the Jupiter. All four missiles used liquid oxygen and kerosene as propellants.

Thor and Jupiter were both single stage ballistic missiles of roughly 1500 miles range. Approximately 90 per cent of the initial weight was propellant compared to 68 per cent for the V-2. Thus, although a substantial performance improvement over the V-2 was attained by the increased specific impulse available from the use of kerosene rather than alcohol as fuel (see table on page 46), the major gain in range from the 200 miles of the V-2 was due to the use of more modern structural techniques resulting in the much higher percentage of missile weight devoted to propellant.

The λ' of a Thor is approximately 0.93, and the vacuum specific impulse of liquid oxygen and kerosene with nozzle expansion ratio of 20 is approximately 275 seconds. This means that the maximum impulsive velocity achievable with a Thor, even with zero payload weight, is about 19,000 feet per second. Thus, the propellant used and structural factors achieved by the Thor still fall short of that required to obtain ICBM range with a single stage vehicle.

In the ICBM case, two different configurations were developed. The Atlas used the same engines as the Thor and Jupiter. It is known as a one-and-a-half stage configuration. At the end of first stage burning, only the booster engines—no tanks—are jettisoned. The Atlas uses two booster engines of 150,000-pound thrust and one sustainer engine of 60,000-pound thrust. The sustainer is ignited at launch along with the other engines. The Titan is a two-stage missile. A similar set of engines was developed for the Titan missile by a different company in order to insure the availability of engines for these high priority programs. Although differing in many design details, the overall performance of the engines is similar.

The gross features of these missiles can be deduced from a knowledge of the specific impulse of the propellants and the velocity requirements from the previous figures. Since the specific impulse was not greatly different from the later solid ballistic missiles discussed earlier, and since roughly equivalent staging was used, the results are similar. The IRBM's were able to carry about four per cent, and the ICBM's about one per cent, of their launch weight as payload.

These missiles occupy a strong position in the thrust into space. They have made large rockets commonplace. The techniques for producing reliable large rocket engines in quantity have been highly developed. The use of cryogenic propellants has become routine. These were the large vehicles under development in the United States when the Space Age dawned in 1957.

Early Space Rockets—A Lesson in Ingenuity

Sputnik I was launched October 4, 1957. The launch date was one day after the fifteenth anniversary of the first successful V-2 shot and about two-and-a-half weeks after the 100th anniversary of Tsiolkovskiy's birth. The first achievement of orbital velocity was an historic event, although, as Figure 3-3 shows, ICBM's are quite close in velocity. The world as a whole was impressed and surprised and the United States reacted violently. As a direct result, a variety of space rocket programs sprang into being.

The versatility of rockets is such that if it is necessary to produce a new launch vehicle in a hurry, almost any type of rocket can be used to boost another type with considerable success. The many examples of this that occurred just after the launching of Sputnik are a display of this phenomenon.

The first American satellite was launched by a Redstone rocket (cryogenic) which contained in its nose two extra stages of solid propellant rockets. This vehicle, which weighed 80,000 pounds, put only a 30-pound weight in orbit—or less than 0.04 per cent of vehicle launch weight. The Vanguard rocket had a cryogenic first stage, a storable liquid second stage, and a solid propellant third stage. The same is true of the Thor/Delta rocket, which put modified Vanguard second and third stages on top of the basic Thor cryogenic stage.

The Thor/Agena is a cryogenic first stage with a storable liquid second stage; the Atlas/Agena, a cryogenic one-and-a-half-stage vehicle with a storable liquid stage on top. The successful Jupiter lunar probe utilized two stages of solid propellant motors on top of the cryogenic Jupiter. This combination, incidentally, used a launch weight of approximately 110,000 pounds to project a 13-pound payload to escape velocity. Thus, the payload was only 0.01 per cent of the vehicle launch weight. That particular shot was probably the most inefficient ever fired and might hold that record indefinitely.

An all-solid-propellant launch vehicle, the Scout, has been used for orbital launches. This vehicle has four stages, and is designed to be a low-cost launcher of relatively lightweight payloads. As the previous discussion of solid propellant ballistic missiles indicates, it is not surprising that solid propellants can be used as satellite launchers.

Probably the best example of how confusing this can become is that although the Thor/Delta is a cryogenic Thor with storable second stage and solid third stage, the Thrust Augmented Thor/Agena uses solid propellants as a half stage to help boost the cryogenic Thor with storable final stage. If this is not complicated enough, a Thrust Augmented Thor/Delta has been used which adds a solid auxiliary boost to the cryogenic Thor with storable upper stage and solid final stage. The essential point to be learned is that rockets are very versatile, and once the basic techniques are in hand, they can be utilized for many purposes. This is why the United States was able to respond to the Sputnik challenge with such a variety of possibilities.

The vehicles mentioned were all used for unmanned launches. In addition, Redstone, Atlas and Titan II represent ballistic missiles built first as weapons which have been successfully adapted for use as manned launch vehicles. A minimum size manned capsule of 1959 design weighs almost 3000 pounds, so that a Redstone could only launch this weight a few hundred miles down range. The Atlas, however, achieved orbital velocity nicely, and a variety of changes and careful procedures implemented especially for manned flight has resulted in 100 per cent reliability as of 1965. It must be realized, however, that only a few flights have occurred, and it seems ridiculous to assume that all of space can be explored without casualties. If so, it will be the only thing the human race has ever done without casualties.

The vehicles mentioned carried the brunt of the early years of the American space program. They have done the job they were called upon to do nicely, although there were vehicle casualties and at times, their payload efficiency was not great. Much of the rest of this book will be devoted to the question of what can be done other than simply to adapt ballistic missiles for space purposes.

4

Lunar and Early Interplanetary Rockets (Velocity up to 100,000 Feet Per Second)

Celestial Mechanics

The calculation of the velocities required to go from the earth to the other bodies in the solar system is an involved process. Velocities beyond local escape (parabolic) velocity result in hyperbolic trajectories and different expressions must be used. In addition, some of the simplicity of the expressions of Chapter 3 were due to the assumption of circular orbits.

The planets of the solar system do not have perfectly circular orbits. Nor do their orbits all lie in the same plane. Furthermore, the axis of rotation of each planet is likely to be in still another plane so that launch and landing points usually are not moving in the desired direction. Since the planets are continually moving around the sun and the natural satellites around the planets, the relative positions and, consequently, velocity requirements to go between the various bodies, are continually changing.

The Multi-Body Problem

The expressions of Chapter 3 were valid for what is known as the two-body problem, namely, that of a space vehicle moving under the influence of a single gravity field. Actually, it was a restricted two-body problem since the mass of the vehicle was assumed to be so small that its effect on the motion of the planet could be ignored. This is a highly justified assumption. When one considers a vehicle going from the Earth to the Moon, a three-body problem is involved. It has never been solved in the completely general case in mathematics. Even the restricted three-body problem, where a particle of negligible mass is moving through the gravity field generated by two other bodies, has only been solved in certain cases. Some many-body (n-body) problems have also been solved for special cases.

The situation is even worse. The Earth-Moon case can only be considered a restricted three-body problem because the effect of the Sun's gravity field does not change substantially between the location of Earth and Moon. For mathematical precision, a rocket flying from Earth to Moon would have to be treated as a restricted four-body problem. In the case of planetary flight, many bodies can be involved.

The mathematical dilemma of the n-body problem can be handled for most cases by using a series of isolated two-body solutions. This is feasible due to the great distances between most of the bodies in the solar system. For instance, one can calculate an Earth-Mars trajectory by first considering only the influence of the Earth. Then, from the point where the Earth's gravitational effect is negligible compared to the Sun's gravity, the vehicle is assumed to move only under the latter's influence. Finally, when close to Mars, it is assumed that only the Mars gravity field affects the motion. By properly matching these solutions, it is possible to understand many celestial mechanics problems.

The following two tables show characteristics of the planets and natural satellites of the solar system. Except for Pluto and Mercury, the planets have nearly circular orbits in planes which are tilted from each other by less than five degrees. The plane of the Earth's orbit is called the ecliptic. Much can be learned by assuming that the planets all have circular, coplanar orbits. This will be assumed for the most part, but occasionally, reference will be made to actual orbits to illustrate the deviations.

The distant planets have extremely long periods about the Sun. The location of the far planets for the remainder of the twentieth century is shown in Figure 4-1. Celestial longitude is measured in the plane of the ecliptic with the Sun at zero

The Planets — Orbital Data

Planet	Symbol	Semi-Major axis A.U.	Peri-helion A.U.	Ap-helion A.U.	Mean Celestial Longitude		
					Of Ascend-ing Node	Of Peri-helion	Of Planet at Epoch 1/1/66
Mercury	☿	.387	.308	.467	47.93°	76.93°	210.29°
Venus	♀	.723	.718	.728	76.38°	131.10°	84.87°
Earth	⊕	1.000	.983	1.017		102.12°	98.89°
Mars	♂	1.524	1.381	1.666	49.30°	335.44°	324.31°
Jupiter	♃	5.203	4.951	5.455	100.11°	13.50°	87.32°
Saturn	♄	9.539	9.008	10.070	113.42°	91.50°	347.57°
Uranus	♅	19.182	18.277	20.087	73.90°	168.65°	166.43°
Neptune	♆	30.058	29.800	30.315	131.40°	53.00°	230.02°
Pluto	♇	39.518	29.692	49.344	109.76°	222.90°	166.77°

The Planets — Orbital Data (Continued)

Planet	Inclination Orbital to Ecliptic	Inclination Equatorial to Orbit	Orbital Velocity about Sun (feet per second)	Period of Revolution Earth Years
Mercury	7.00°		157,000	0.240
Venus	3.39°		114,800	0.615
Earth		23.45°	97,600	1.0
Mars	1.85°	25.20°	79,100	1.881
Jupiter	1.31°	3.12°	42,800	11.86
Saturn	2.49°	26.75°	31,600	29.46
Uranus	0.77°	97.98°	22,200	84.02
Neptune	1.77°	29°	17,800	164.78
Pluto	17.13°		15,500	248.4

Solar System Data

SOLAR BODY	Semi-Major Axis A.U.	Semi-Major Axis Planet Radius = 1	Planet Mean Diameter Miles	Planet Mean Diameter ⊕ = 1	Mass Ratio ⊕ = 1	Surface Gravity ⊕ = 1	Period About Primary Earth Days	Escape Velocity (feet per second)
SUN			869,000	101	333,500	27.7		2,020,000
MERCURY	.387		3,010	.38	.053	.367	88	13,700
VENUS	.723		7,710	.97	.815	.862	225	33,600
EARTH	1.00		7,920	1.00	1.00	1.00	365	36,700
Moon		60.27	2,160	.272	.0123	.166	27.32	7,800
MARS	1.524		4,220	.53	.107	.376	687	16,400
Phobos		2.775	10				.32	
Deimos		6.919	5				1.26	
ASTEROIDS								
Ceres	2.767		460				1681	
Pallas	2.767		300				1684	
Juno	2.670		120				1594	
Vesta	2.361		240				1325	
JUPITER	5.203		88,600	11.20	318.0	2.54	4333	196,000
V		2.539	100				.50	
Io		5.905	2,060	.26	.0132	.195	1.77	8,250
Europa		9.396	1,790	.23	.0080	.156	3.55	6,900
Ganymede		14.99	3,070	.39	.0256	.170	7.15	9,430
Callisto		26.36	2,910	.37	.0151	.112	16.69	7,450
VI		160.1	75				250.6	
VII		164.4	25				259.8	
X		164	12				260	
XII		290	12				625	
XI		313	15				696	
VIII		326	25				739	
IX		332	14				755	

Solar System Data (Continued)

SOLAR BODY	Semi-Major Axis A.U.	Semi-Major Axis Planet Radius = 1	Planet Mean Diameter Miles	Planet Mean Diameter ⊕ = 1	Mass Ratio ⊕ = 1	Surface Gravity ⊕ = 1	Period About Primary Earth Days	Escape Velocity (feet per second)
SATURN	9.539		75,000	9.47	95.22	1.06	10,759	116,000
Mimas		3.111	300				.94	
Enceladus		3.991	350				1.37	
Tethys		4.939	750	.09	.00011	.013	1.89	1,310
Dione		6.327	800	.10	.00017	.017	2.74	1,540
Rhea		8.835	1,100	.14	.00039	.020	4.52	1,950
Titan		20.48	3,100	.39	.0230	.150	15.95	8,900
Hyperon		24.83	250				21.28	
Japetus		59.67	750				79.33	
Phoebe		216.8	200				550	
URANUS	19.182		29,600	3.74	14.55	1.04	30,687	72,400
Miranda		5.494	350				1.41	
Ariel		8.079	350				2.52	
Ubriel		11.25	250				4.14	
Titania		18.46	600				8.71	
Oberon		24.69	500				13.46	
NEPTUNE	30.058		27,800	3.50	17.23	1.41	60,184	81,600
Triton		15.85	2,500	.31	.0252	.256	5.88	10,400
Nereid		249.5	200				500	
PLUTO	39.518		9,000?	1.1?	.9	.705	90,700	32,700

Surface gravity and escape velocity calculated from diameter and mass ratio.

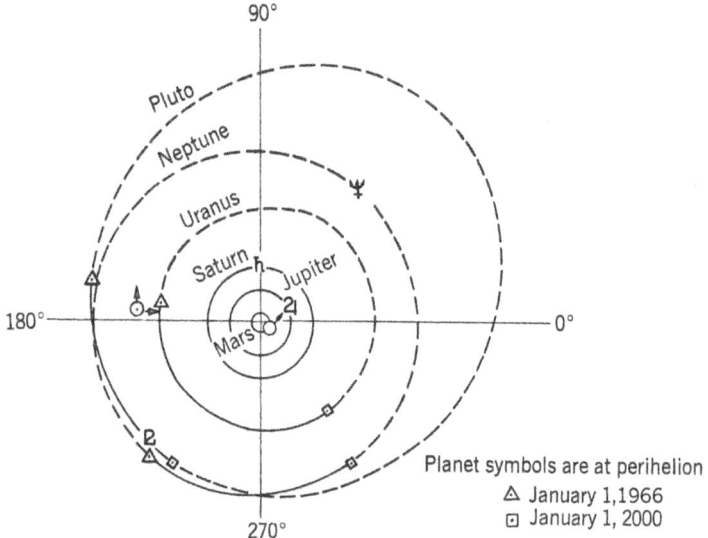

Fig. 4-1. The outer solar system.

degrees with respect to the earth at the vernal equinox (about March 21). Pluto is now relatively close, and moving toward perihelion about 1990. Pluto will be closer than Neptune from about 1980 to 1997. Unfortunately, however, Pluto is close to its highest inclination now.

Hyperbolic Excess Velocity

Escape velocity was derived in Chapter 3 by calculating the initial kinetic energy which was just equal to the maximum potential energy. In that case, the vehicle had zero velocity at an infinitely great distance. If more than escape velocity is imparted at launch, then at an infinite distance, the vehicle will still possess energy, since only an amount equal to the maximum gravitational potential can be removed by the gravity field.

The velocity remaining at an infinite distance is known as the hyperbolic excess velocity since trajectories with these characteristics are hyperbolas. The Vis-Viva Law can be expressed in terms of the hyperbolic excess velocity since one obtains from Equation 3-11, for $r \to \infty$:

$$V_\infty^2 = -\frac{g_0 R^2}{a} \qquad (4\text{-}1)$$

where V_∞ = hyperbolic excess velocity.

Note that a, which represents the semi-major axis for elliptical orbits, is a negative number for hyperbolic orbits. Equation 4-1 may then be substituted back into Equation 3-11 to give, with the aid of Equation 3-7:

$$V^2 = V_E^2 + V_\infty^2 \qquad (4\text{-}2)$$

or:

$$V_\infty = \sqrt{V^2 - V_E^2} \qquad (4\text{-}3)$$

Equation 4-2 gives a simple but useful expression for the velocity of a vehicle at a great distance from a gravity source. To calculate the launch velocity required to reach the orbit of Mars, for instance, Equation 3-10 may be used to calculate the "perigee" velocity needed. This gives the heliocentric velocity (heliocentric velocity means that with respect to the Sun as opposed to geocentric, with respect to Earth, selenocentric with respect to the Moon, etc.) at perihelion, which is made up of the earth's orbital velocity around the sun and the vehicle's velocity with respect to the Earth. Equation 4-2 provides a simple means of calculating the latter.

Equation 4-2 also provides further insight into the interaction between space vehicles and gravity fields. Although the same amount of energy is always extracted from the vehicle by the gravity field, the fact that kinetic energy varies as the square of the velocity means that, at higher velocities, this amount of energy represents an increasingly smaller part of the total velocity. Figure 4-2 shows the hyperbolic excess velocity as a function of initial and escape velocity. This curve illustrates the importance of modest increases in initial velocity. If only 20 per cent more than escape velocity is supplied, then over 50 per cent of all velocity supplied remains at great distance.

One effect of Equation 4-3 is that high velocity rockets that cross gravity fields quickly are far more effective at covering the large distances of space than might be thought at first from Hohmann Transfer calculations. Indeed, the effect is compounded by the rapid traversing of both the Earth and the Sun's gravity fields. Figure 4-3 shows both Earth and solar system hyperbolic excess velocity as a function of Earth launch velocity. The rapid gain with launch energies beyond minimum is so great that at an Earth launch velocity of 67,000 feet per second, the solar system hyperbolic excess velocity is also 67,000 feet per second.

If one were to launch toward a distant star, the result of launching from the earth at 67,000 feet per second would be the same as by launching at the same velocity from deep in space away from all effects of Earth and Sun gravity. At launch velocities beyond 67,000 feet per second, the solar hyperbolic excess exceeds the Earth launch velocity. This apparently free velocity is due to the fact that the Earth is

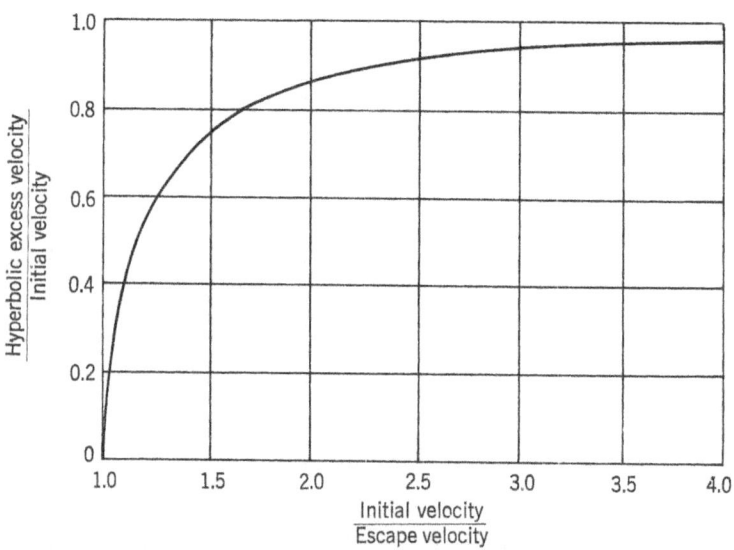

Fig. 4-2.   Hyperbolic excess velocity.

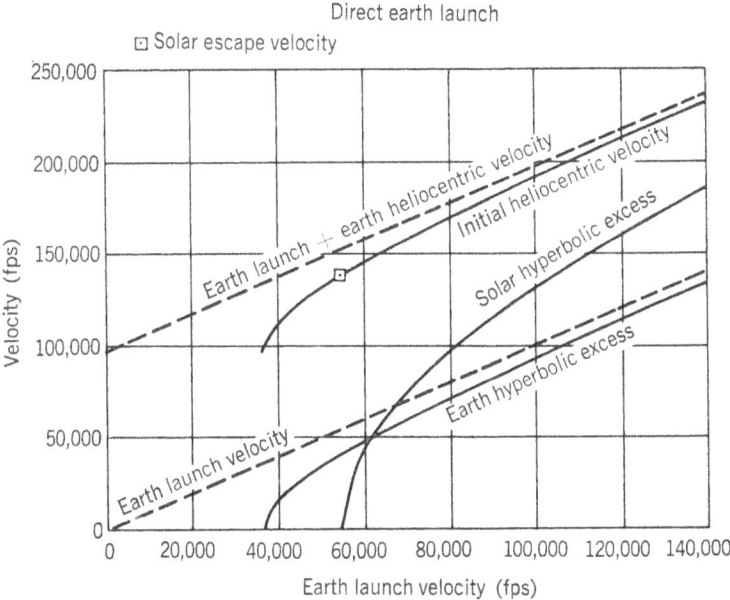

Fig. 4·3. Solar system hyperbolic excess velocity.

already on orbit around the Sun and the Earth's heliocentric orbital velocity of 96,700 feet per second is being partially utilized. Whatever mechanism placed the Earth on its current orbit has supplied energy which can now be utilized.

Braking within Gravity Fields

When a planet is approached, the vehicle is accelerated by the gravity field. The closer the vehicle comes to the planet, the greater will be the kinetic energy imparted to the rocket fuel by the planet's gravity. Thus, a greater amount of kinetic energy will be dissipated by the rocket when braking is applied. For the same reason that launch velocity inputs should be made at the lowest altitude possible, all braking velocity removals should be done as close as possible to the center of gravitation, whether the vehicle is landing on the planet or

braking into an orbit around it. This applies equally to elliptical, parabolic, or hyperbolic trajectories. For hyperbolic
approach, Equation 4-2 gives the braking velocity required for
a given hyperbolic excess, since it is the same as the initial
velocity.

Atmospheric Braking

Extensive calculations have been made of atmospheric
braking and entry for the planet Earth and substantial calculations have been performed for both Mars and Venus. A
very rough rule of thumb from this work is that 1.5 times the
planetary escape velocity could be braked aerodynamically by
a lifting body if a planetary landing were being made and
that 0.5 times escape velocity could be braked in an atmospheric pass while establishing a satellite orbit. For many cases
of interest, the atmospheres of the planets Earth, Mars, and
Venus, may be capable of supplying all necessary braking
velocity requirements, although the other two minor planets
have tricky braking problems. The Mercury braking requirement is large because of the high orbital velocities close to
the Sun. The Pluto requirement is large if high launch velocities are used to decrease travel times. By coincidence, both
tend to be about 50,000 feet per second. The atmosphere of
Mercury is estimated to be as dense at the surface as that of
Earth or Mars at about 150,000 feet. Due to low gravity, the
Martian atmosphere is likely deeper than Earth's and more
effective for braking at high altitudes. Since Mercury and Mars
are small, the atmosphere of Mercury should be about as
effective as that of Mars except for touchdown requirements.
Some aerodynamic braking may thus be possible at Mercury.
Pluto is thought to be slightly larger than Earth, and might
have an atmosphere which has remained undetected due to
the great distance of observation. On the other hand, the very
low temperature of Pluto should have frozen out the normal
atmospheric gases, and the atmosphere of such a cold planet
may be vastly different from Earth's. Thus, atmospheric
braking at Pluto is a complete unknown at present.

The large planets have extensive atmospheres, but their surfaces appear extremely forbidding. It is even doubtful if they have solid surfaces. It will be seen that braking velocity removals of 10,000 feet per second cover many of the requirements at the major planets. A velocity decrease of only 10,000 feet per second in such large atmospheres, even if exact knowledge of the atmospheric composition is not known, would seem easy. In this case, however, the effect of the planetary gravity field is a disadvantage. The high approach velocities, which impart useful kinetic energy to rocket fuel, simply mean that larger energies must be dissipated for a given velocity decrease when aerodynamic braking is used. Atmospheric landing on the "surfaces" of the major planets may be feasible, but will not be considered further here.

Lunar Velocity Requirements

The velocity requirements of a lunar mission are easily estimated. The distance from the earth to the moon averages about 60.5 earth radii (the moon has a slightly elliptical orbit) and Equation 3-10 gives 99.2 per cent of escape velocity or 36,400 feet per second as launch velocity. The velocity at apogee of the trajectory will be 600 feet per second (see Equation 3-9) and since the moon's orbital velocity is about 3350 feet per second, the hyperbolic excess velocity with respect to the moon would be the difference of these two, or 2750 feet per second. Since the moon's escape velocity is 7800 feet per second, Equation 4-2 may be used to calculate 8300 feet per second as a braking velocity required to land on the moon. Thus, as an absolute minimum, twice 8300 feet per second must be added to 36,400 feet per second, or 53,000 feet per second must be supplied.

If atmospheric braking is used on return to earth, then the addition of roughly 4000 feet per second for drag and gravity losses, and 1000 feet per second for guidance correction capability is adequate. In this case, travel time to the moon each way will be approximately four days. It is possible to reduce this time to half by adding less than 500 feet per second additional launch velocity. There is little incentive to do this, however, since the four-day, one-way, flight time is no unusual burden as yet. Furthermore, the braking requirements at the moon increase rapidly for small increases in launch velocity. This would be expected from Figure 4-2.

Thus, once again, the velocity required can be bracketed by relatively elementary calculations. When planning flight operations, extensive calculations are made to define precisely velocity requirements, including three-body effects, mechanize guidance equations, and provide special flight plans so that it is easier to abort the operation and successfully return the expedition to earth in case of emergencies. Special requirements to reach various areas of the moon on arrival, and return to selected spots on earth, also make the actual calculations complicated.

Planetary Hohmann Transfers

The planets are sufficiently far away from the Earth that it is customary to use a different measure of their distance. Since the Moon is only about 240,000 miles away from Earth, it is relatively convenient to use miles. When measuring planetary distances, however, the astronomical unit (A.U.) is used. It is defined as the average distance between the Earth and the Sun, and is approximately 92.96 million miles. An idea of the size of the solar system may be obtained by realizing that Pluto at aphelion is 49.3 astronomical units from the sun, or about 4.6 billion miles. It takes almost seven hours for light from the Sun to reach Pluto at aphelion.

It is natural to try to use minimum energy trajectories throughout the solar system. In fact, preoccupation with the apparent extreme difficulty of building high velocity rockets led Hohmann to derive his minimum energy transfer system. Hohmann transfers between planets are calculated as described in Chapter 3, using heliocentric and planetary orbital velocities with Equation 4-2 used to convert planetary hyperbolic excess to launch or braking velocity. The period of the transfer orbit is given by Equation 3-13.

Figures 4-4 and 4-5 give the result of Hohmann Transfer calculations between the Earth and the other planets in the solar system, including requirements to establish both highly elliptical and low circular orbits around the various planets. Essentially the same velocity capability as required of the lunar and return rocket (58,000 feet per second) enables flight anywhere in the solar system. Unfortunately, travel times throughout most of the system are excessive.

In addition to basically long flight times, the use of Hohmann Transfers between the planets has another undesirable effect. Since the planets rotate around the Sun at different rates, they constantly change their positions with respect to each other. Consequently, after arrival at a planet, it is necessary to wait until conditions are right for return. The term "launch window" is widely applied to that period of time when celestial bodies are in proper position for launch.

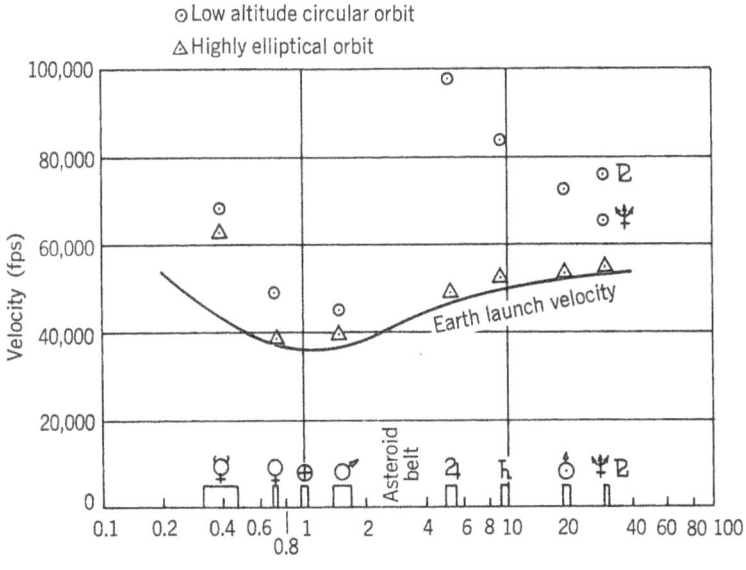

Fig. 4-4.   Hohmann transfer velocities.

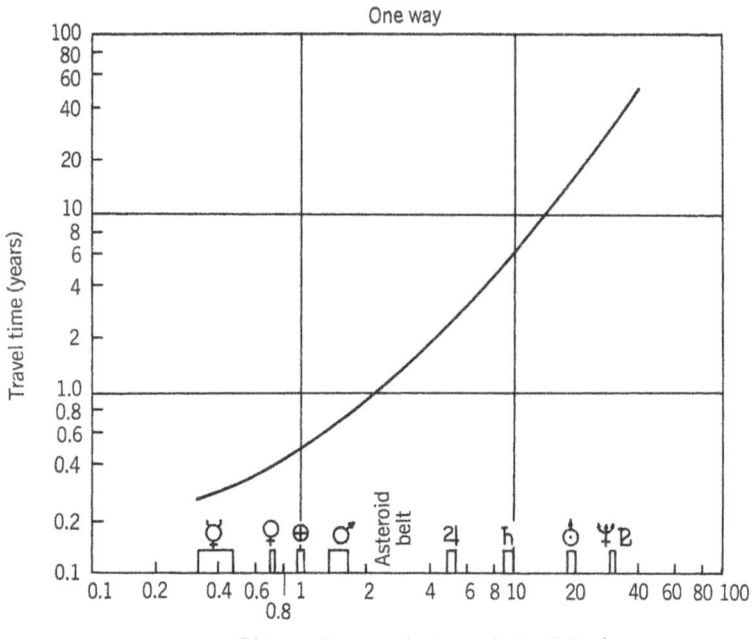

Fig. 4-5.   Hohmann transfer travel time.

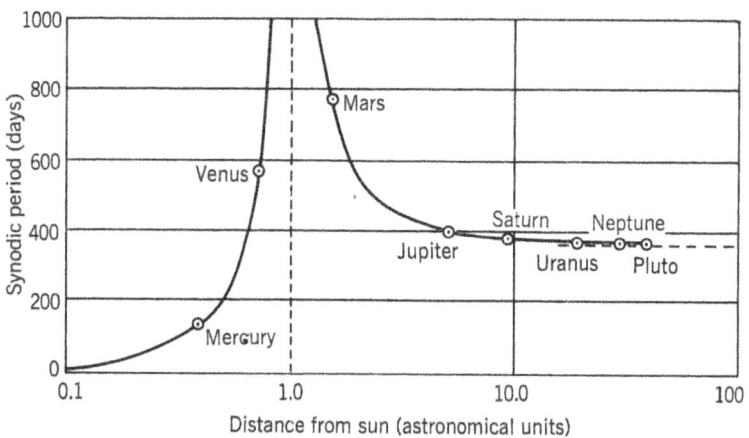

Fig. 4-6.   Synodic period of planets.

The synodic period of two planets is the term applied to the time interval between successive conjunctions. Conjunction occurs when the planets are on exactly opposite sides of the Sun. The synodic period is:

$$P_s = \frac{1}{\dfrac{1}{P_1} - \dfrac{1}{P_2}} \qquad (4\text{-}4)$$

where P_s = synodic period; P_1 = orbital period of planet 1 about the Sun; and P_2 = orbital period of planet 2 about the Sun.

Figure 4-6 gives the synodic period of Earth with the rest of the planets in the solar system. Because Mars and Venus are close to the Earth, they have long synodic periods. The far planets are moving around the Sun so slowly that they act almost as fixed points, their synodic period approaching the Earth's orbital period of one year. For planets closer to the Sun than Venus, the synodic period is short since the orbital period of the inner planet is small.

For Mars, the waiting time after arrival until the next launch opportunity is 15 months. This, combined with the one way Hohmann travel time of nine months, means that an absolutely minimum energy Hohmann trip to Mars and back would require 33 months. It is not surprising that one is interested in higher velocities for interplanetary missions. As we pursue faster transfers, however, the simplicity of Hohmann calculations disappears. The expressions given so far are a great aid in understanding basic physical effects, but the varieties of trajectories and general complexity of the solar system force reliance on curves of the results of extensive calculations.

Multi-Impulse Transfers

The Hohmann Transfer to a distant orbit is the minimum energy possible. It is also minimum velocity if we only utilize "two impulse trajectories," namely, those with one velocity input at periapsis and one at apoapsis injecting directly into the final orbit. Three impulses can be used by supplying more initial velocity so that the rocket goes to a higher apoapsis than the final orbit. At that point, an impulse is used to raise the periapsis to the final radius desired and then a third velocity input is made when reaching subsequent periapsis. Hohmann did not check this possibility.

It can be shown that for transfers to orbits greater than 5.8 times initial radius, this particular maneuver will create the desired orbit for less velocity, even though the energy required is greater. The flight times are much longer. The gains by this maneuver are small, and, as of 1965, only two impulse trajectories have been used.

The gains by using multi-impulse trajectories can be large when the objective is to reach a low periapsis starting from a high orbit. An example is a solar probe vehicle which is trying to reach as close to the sun as possible when launched from the earth. The Hohmann Transfer method would utilize one impulse to nullify most of earth orbital velocity to reach the sun. Instead, one could use two impulses, launching outward from earth to great distance and then decreasing velocity at aphelion. Figure 4-7 shows the velocity required to perform such maneuvers. When trying to reach final radius less than 0.2 initial radius, substantial gains can be made by the use of two impulses.

Figure 4-7 shows a case where velocity inputs should be made at great distances. Previously, it has been emphasized that velocity changes should be made as deep in the gravity field as possible in order to take advantage of the high initial velocity of the vehicle. This previous conclusion clearly only applies if the initial velocity is in a helpful direction, as when one is adding or subtracting energy in the direction of the trajectory. If the objectve is to make major changes in the tilt of the trajectory, a large velocity vector is a detriment. At apoapsis, a vehicle will be moving slower than at periapsis and inclination changes from the basic trajectory can be made with smaller velocity inputs. In the extreme case, at an infinite distance, all elliptical orbits have zero velocity and the inclination of the orbit can be changed to any desired value with zero

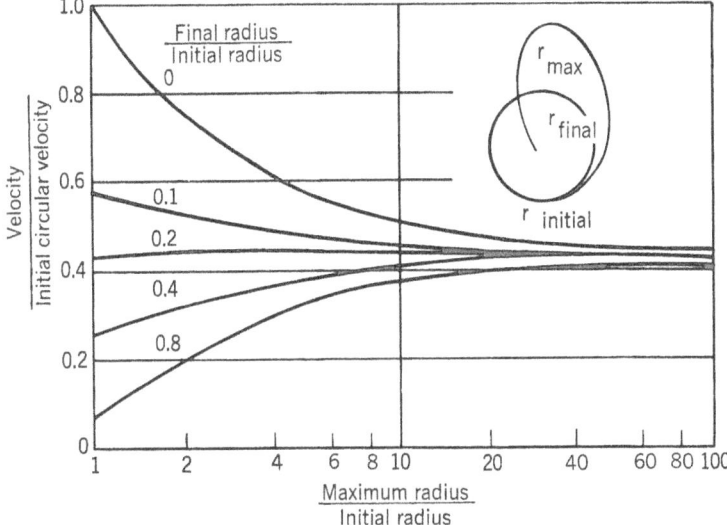

Fig. 4-7.   Solar probe-type missions with two-impulse transfers.

velocity input. In addition, the periapsis can be placed at any desired radius, which is why the curves of Figure 4-7 coincide at large values of maximum radius.

Use of Planetary Energy

It is possible to make use of the energy of planets to reduce the rocket energy required for interplanetary travel. This is done by flying close enough to a planet to use its gravity field to deflect the vehicle so as to put in or extract energy from the trajectory. This mechanism can be understood by reference to Figure 4-8, for the net result is the same as if the vehicle had been bounced off the planet.

If two bodies collide, a perfectly elastic impact is said to have occurred if no energy is dissipated in the collision. This would occur if a perfect spring were to absorb the energy of impact and then re-transmit this energy to the two bodies by springing back to its original length. Perfect springs are hard

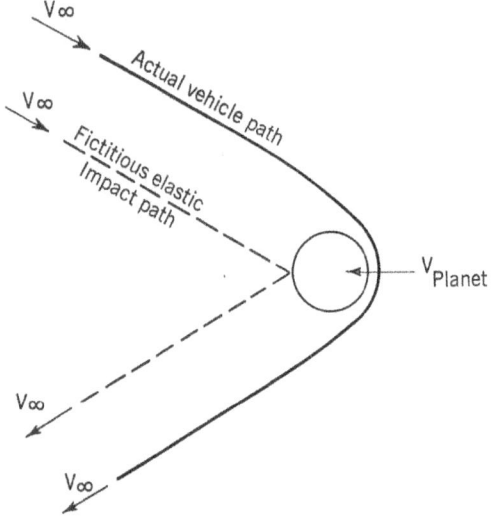

Fig. 4-8.   Elastic impact analogy for the use of planetary energy.

to find. The mechanism by which a vehicle swings around a
planet, if viewed from a distance, appears like a perfectly
elastic impact even though the vehicle never hits the planet.
The planet is approached with a given hyperbolic excess
velocity and when the vehicle departs from the influence of
the planet's gravity field, it has the same hyperbolic excess
but in a different direction. No energy is dissipated in the
process.

   The role of the planet's gravity field in this process is similar to that of a perfect spring, since it absorbs the energy of
the incoming vehicle and re-transmits it with perfect efficiency. It does this in a manner somewhat like lassoing the
flying body, pulling on the rope to swing it around, and then
letting go. Since there is finite energy in the planet's gravitational field, there are limitations upon the angle through which
the trajectory can be turned. The simple relation between
turning angle, hyperbolic excess velocity, and energy of the
gravity field at closest approach is shown in Figure 4-9.

   In describing the swing-around process so far, it is not clear how energy can be obtained since there is no change in vehicle energy with respect to the planet before and after. One cannot take energy from the planet's gravity field, since the field contains the same amount of energy before and after. It could only be changed by removing mass from the central gravitating body. The energy which can be gained comes from the orbital motion of the planet.

By equating the linear momentum of vehicle and planet before and after an elastic head-on impact, one obtains (exactly analogous to the first equation in this book):

$$w_V(V_\infty + V_{P_I}) + w_V(V_\infty - V_{P_F}) = w_P(V_{P_I} - V_{P_F}) \quad (4\text{-}5)$$

where w_V = weight of vehicle in pounds; w_P = weight of planet in pounds; V_{P_I} = initial velocity of planet in feet per second; and V_{P_F} = final velocity of planet in feet per second. If the vehicle is of negligible size with respect to the planet, the planet will lose almost no velocity, and Equation 4-5 becomes:

$$w_V(2V_\infty) = w_P(\Delta V_P) \quad (4\text{-}6)$$

where ΔV_P = planet velocity loss in feet per second. In this maximum case, the magnitude of vehicle velocity in heliocentric coordinates which gives the heliocentric energy change

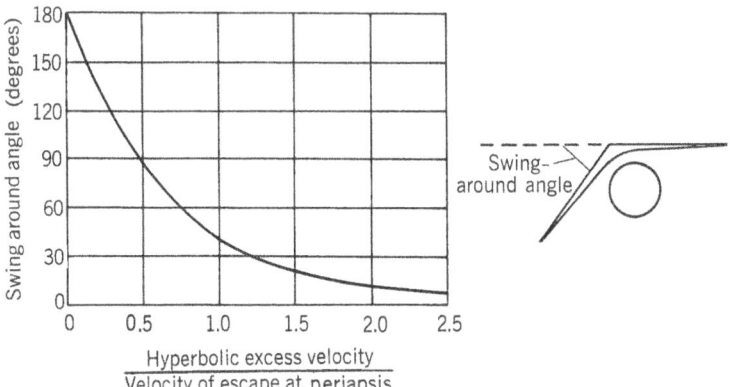

Fig. 4-9.   Planetary swing-around angle.

is twice the planet's orbital velocity. If the "impact" is not head-on, less than this value will be obtained. Were the planet not moving, the vehicle velocity direction would change but its heliocentric energy would not.

If one were to fly enough vehicles by a planet in this manner, it is at least theoretically possible to extract enough energy from the planet's orbit so that it would spiral into the Sun. There is relatively little danger of this, however, since Equation 4-6 shows the planet velocity decrease to be reduced by the ratio of vehicle to planet weight. Even on a planet as small as Earth, if a million-pound vehicle performed a billion flybys with 10,000 feet per second hyperbolic excess velocity, the Earth would lose only one millionth of a foot per second velocity. We are dealing with the same relations which underlie the rocket equation and a large fraction of the Earth's weight would have to be utilized in such a process to make significant changes in the Earth's velocity. In addition, if the technique were widely used, it would likely be used as often to decrease vehicle energy as to increase it. Hence, the planet would be nudged forward as often as backward and the net effect would approximately balance.

The use of planetary flybys further complicates an already difficult celestial mechanics situation by adding one more variable. It is probable that as of 1965, all the useful possibilities have not been unearthed. In many cases, the benefits to be gained are not so much in the area of reducing velocities below that of a Hohmann Transfer but are more likely to permit operation for Hohmann class velocities during other times of the synodic period.

The Russian Lunik III used a close flyby of the moon for trajectory control in 1958. By this means, the probe was returned in the proper position for data read-out, and its perigee was raised to increase orbital life time.

With respect to the use of planetary energy, the four inner planets behave distinctly differently from the four major planets. The ratio of the escape velocity of the inner planets to their orbital velocity about the Sun (see table on page 82), varies from .087 to .376. In all cases, it is much less than 1.00.

Hence, a swingby of any of these planets will not be capable of making gross changes in velocities of magnitude equal to the planet's orbital velocity (see Figure 4-9). Gross changes in the trajectory cannot be made, but very useful trajectory deflections are possible.

If we launch from Earth to Mars, for instance, only a limited angular spread of trajectories is possible. If Venus is in the path of this trajectory fan, then by varying the closeness of approach to Venus, the trajectories can be fanned out further in either direction and Mars can be intercepted over a wider range of orbital positions. Although the launch velocities for Earth-Mars operations are not decreased significantly from Hohmann values by this process, it is typically possible to find Earth/Mars launch windows for half of the Earth/Venus launch windows. Since Venus launch windows are more frequent than those of Mars and rarely occur at the same time, this represents almost a doubling of the available launch windows to Mars.

For the major planets, the ratio of planetary escape velocity to orbital velocity about the Sun ranges from 3.25 to 4.57. All of these planets could swing velocities equal to their own orbital velocity well beyond 90 degrees. If a vehicle approaches with heliocentric velocity not greatly different from the planet's orbital velocity, gross changes in the heliocentric trajectory can be made. The vehicle can be deflected into the Sun or completely out of the solar system. If one had enough patience and good enough guidance systems, a probe could be deflected from Jupiter around the Sun in such a way as to travel to Saturn. A proper swingby of Saturn could throw it back around the Sun to one of the other major planets. This presumably could be carried on indefinitely or the vehicle could be flipped out of the solar system. It is unfortunate we do not have a large planet as close as either Venus or Mars, for the trajectories possible by close approach to such a planet could be quite spectacular.

The major planets may be used to deflect trajectories to close approaches of the Sun and out-of-ecliptic missions. The velocity requirements are sometimes greatly reduced since the

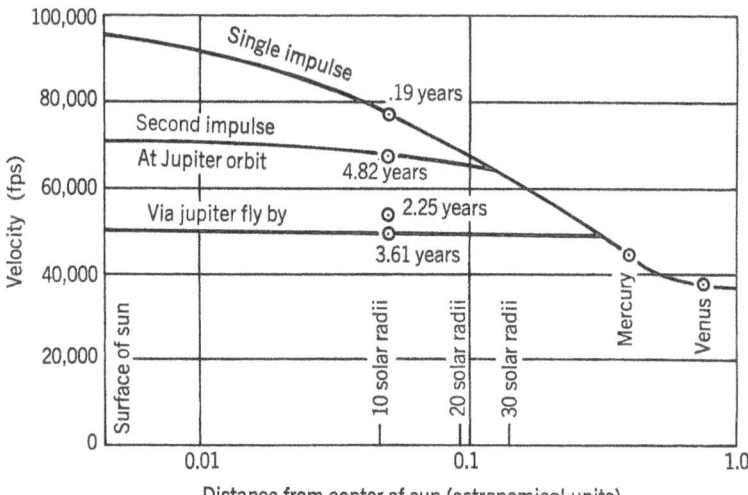

Fig. 4-10.   Solar probe velocity requirements.

gains shown in Figure 4-7 by applying impulse at aphelion
are greater if a planet supplies the aphelion velocity change.

The use of Jupiter is particularly effective. In addition to a
large gravity field adequate for either solar probe or out-of-
ecliptic missions, Jupiter is close enough to the Sun that the
flight time increases for Jupiter flyby trajectories compared
to direct flights are not excessive.

The use of a Jupiter flyby for solar probe missions is shown
in Figure 4-10. The velocity required to come as close to the
Sun as desired is 54,000 feet per second if about 2.25 years' flight
time can be tolerated as compared to a velocity of 80,000 feet
per second required to approch to only 10 solar radii by conventional trajectories. The use of a second rocket impulse at
the orbit of Jupiter is also shown in Figure 4-10.

An even more startling result occurs for out-of-ecliptic
trajectories as shown in Figure 4-11. To launch directly from
the Earth to 90 degrees out-of-ecliptic and go over (or under)
the Sun with a closest approach of one astronomical unit requires 140,000 feet per second. The same maneuver making
use of a Jupiter flyby requires only 52,000 feet per second.

Likewise, the requirement for going 90 degrees out-of-ecliptic and making the closest possible approach to the Sun is reduced from 105,000 feet per second to 50,000 feet per second. The minimum energy curves of Figure 4-11 represent varying degrees of closest approach to the Sun. At low out-of-ecliptic angles, the probes stay essentially at one astronomical unit from the Sun; for out-of-ecliptic angles approaching 90 degrees, the minimum energy trajectories approach closely to the Sun.

The discussion in this section on the use of planetary orbital energy may seem a rather long celestial mechanics discourse to appear in a text on space propulsion. It is a prime example, however, of the inseparability of these subjects. Planetary energy, which might be considered a form of "gravity propulsion" is free energy, available in reliable form for the price of clever guidance. The good propulsion engineer will use whichever form of propulsion is most appropriate and should never

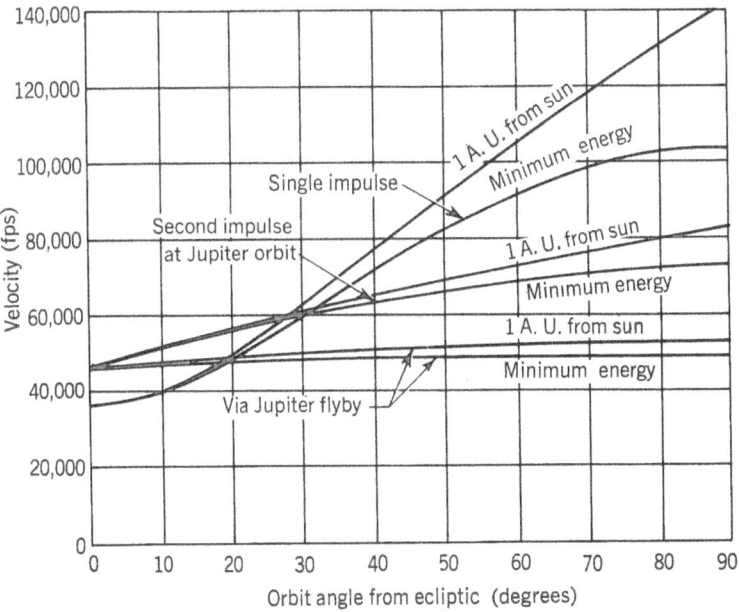

Fig. 4-11.   Out-of-ecliptic velocity requirements.

build needlessly large rockets if small rockets supplemented by planetary energy are equally effective.

Escape from the Solar System

If one launches from the Earth in the direction of the Earth's rotation around the Sun, then as explained earlier in this chapter, and as plotted in Figure 4-3, 54,500 feet per second is required to escape from the solar system. This is commonly referred to as solar system escape velocity and is frequently thought to be the minimum velocity required to escape from the solar system. It is, if only a direct launch from Earth is utilized.

If one launches to Jupiter, a velocity of about 47,000 feet per second will be required. When the vehicle arrives, Jupiter is capable of deflecting it out of the solar system. Hence, it is possible to escape from the solar system with only 47,000 feet per second launch velocity rather than 54,500 feet per second. Interestingly enough, the flight time to Pluto this century using this lower launch velocity is only 18 years compared to 30 years without the aid of the Jupiter gravity field. In this case, Jupiter is unusually effective, since it is also able to deflect the probe to the high inclination of Pluto.

One could imagine other flight procedures: for instance, a flyby of Venus en route to Jupiter, a flyby of Mars en route to Jupiter, or even a Venus-to-Mars-to-Jupiter possibility. As long as the vehicle reaches Jupiter, it can almost certainly be deflected out of the solar system there. One might even consider using a flyby of the Moon to aid in getting some of the velocity needed to go to Venus or Mars. None of these possibilities has yet been shown to give a lower escape velocity than launching directly from Earth to Jupiter. The only possibility appears to be the use of a multiple Mars swingby, each time turning the trajectory to one which will intercept Mars after several solar orbits, and eventually turning into one which reached Jupiter. It might be possible to escape from the solar system, if flight time were no object, for about 40,000 feet per second Earth launch velocity by this means.

Minimum Energy Fast Transfers

The Earth launch velocity required to reach the outer solar system is given for various travel times in Figure 4-12. Large decreases in travel time result for relatively modest increases in launch velocity. For example, minimum energy flight to Uranus requires 52,000 feet per second velocity and takes 16 years (see Figures 4-4 and 4-5). But this time could be reduced to four years by the use of 63,000 feet per second launch velocity.

It is desirable to reduce travel time for several reasons. In manned expeditions, the awkwardness of extremely long duration expeditions is obvious. In unmanned probes, reliability of equipment is the most obvious reason. In addition, however, if one thinks of flight times of 16 years (or 30 years to Pluto), one must consider the development plus vehicle travel time of the system under discussion compared to whatever new

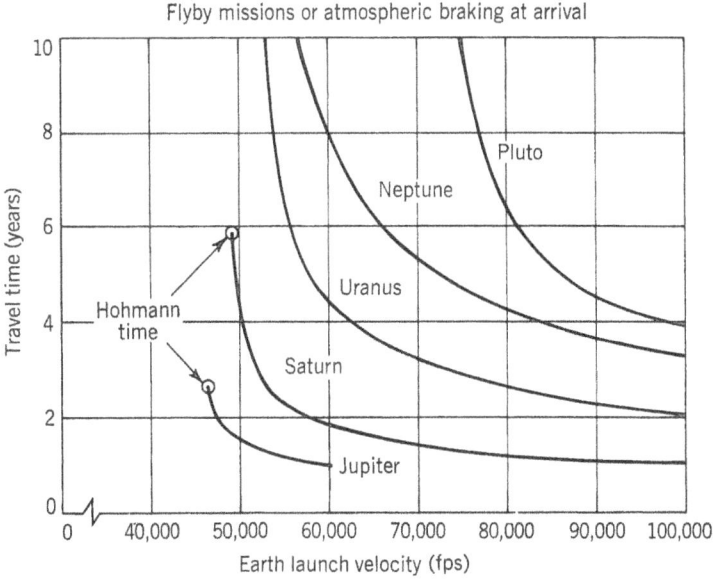

Fig. 4-12.   Solar system travel times.

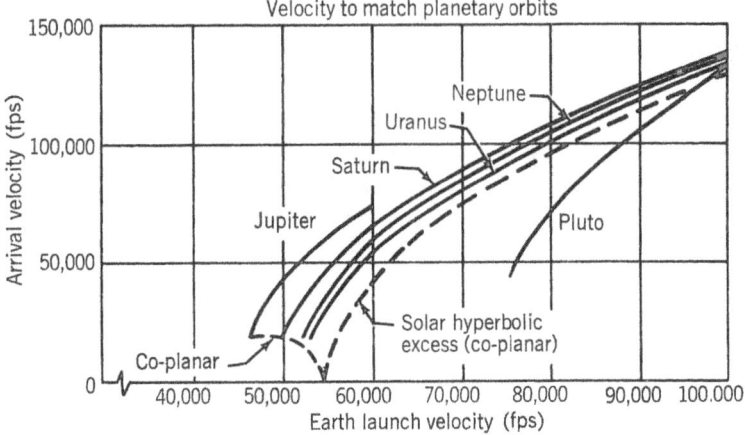

Fig. 4-13.　Planetary arrival velocities.

system will replace it. This probably places a restriction on maximum flight time of about 10 years.

Planetary arrival velocities (hyperbolic excess) corresponding to the launch conditions of Figure 4-12 are shown in Figure 4-13. The curve for Pluto does not fit the pattern of the other planets due to the high Earth launch velocity required by Pluto's high inclination.

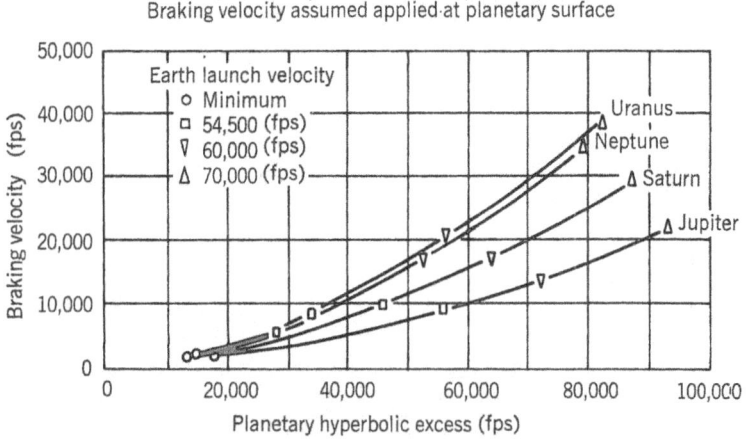

Fig. 4-14.　Planetary capture velocities.

Upon arrival at another planet, braking velocity will be applied except in the case of a flyby mission. It may be desired to land on the planet, go into orbit about the planet for surveillance, or go into an orbit as part of the process of landing on one of the planet's natural satellites. The large planets—Jupiter, Saturn, Uranus, and Neptune—have large gravity fields to aid in the efficiency of braking. Figure 4-14 shows the actual braking velocity which must be applied at each of the planets to go into an elliptical orbit with apoapsis of 100 planetary radii. The very strong attenuation of the arrival velocities of Figure 4-13 by the planetary gravity fields is evident.

Landing on Natural Satellites

After the vehicle is established in a capture orbit, its manner of transfer to final circular orbit should depend on the radius of final orbit. If the final orbit is less than about five planetary radii, the vehicle should atmospherically brake (if possible) an amount of velocity upon reaching next periapsis so that the subsequent apoapsis would be at the desired orbital radius. The velocity required to establish orbit is added at subsequent apoapsis. If the final orbit is greater than five planetary radii, the vehicle should add an amount of velocity at initial apoapsis to raise the periapsis to the desired radius, then subtract the necessary velocity to establish orbit upon reaching periapsis. The velocities to be added or subtracted at final orbit injection for the major planets are shown in the curves of Figure 4-15.

Velocity requirements to match orbital velocities and land on various satellites also are shown in Figure 4-15 under the assumption the orbit-matching maneuver takes place in close proximity to the satellite involved. Although 16,000 feet per second would permit landing on almost all satellites of all planets, the Galilean satellites of Jupiter could not be reached. Eighteen-thousand feet per second would give the ability to reach Callisto. The hardest to reach—Io, would require 26,000 feet per second.

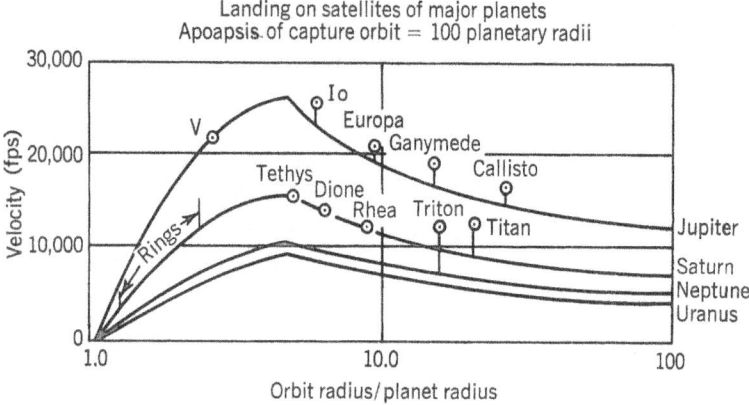

Fig. 4-15.   Payload velocity requirements.

Asteroids and Comets

Asteroids are relatively small objects which exist mostly in a belt between Mars and Jupiter. Thousands have been discovered. Their total mass has been estimated to be about three per cent of our Moon's mass. The four largest asteroids are listed in the table on page 82. If the asteroid belt does represent a shattered and hence pre-mined planet, as many people believe, it was a very small one. Even so, if the material in the asteroid belt were distributed uniformly over the land surface of Earth, it would form a layer about 2.75 miles deep. That is quite a mining operation, spectacularly so, if the material turns out to be different from Earth.

If iron meteorites do represent material from the core of a planet which existed there, an iron asteroid one mile in diameter containing nine per cent nickel (typical of meteorite percentages) would supply the current world production rate (400,000 tons in 1962) for over 4000 years. Since iron meteorites contain about 0.5 per cent cobalt, the same asteroid might supply current cobalt needs for even longer. The asteroids are typical of space objects which would be of great interest, if transportation problems could be solved.

Selected Comets

Comets	Perihelion Distance (A.U.)	Aphelion Distance (A.U.)	Period (Years)	Perihelion Passage Time
Encke	0.339	4.09	3.30	1967 Sep 12
Grigg-Skjellerup	0.855	4.88	4.90	1966 Nov 24
Tempel (2)	1.369	4.68	5.27	1967 Aug 18
Forbes	1.545	5.36	6.42	1967 Dec 21
Wolf-Harrington	1.604	5.37	6.51	1971 Aug 19
D'Arrest	1.378	5.73	6.70	1967 Jun 17
Reinmuth (2)	1.933	5.18	6.71	1967 Aug 11
Borrelly	1.452	5.88	7.02	1967 Jun 20
Faye	1.652	5.95	7.41	1969 Dec 29
Comas-Sola	1.777	6.61	8.59	1969 Nov 4
NeujMin (1)	1.547	12.17	17.95	1966 Dec 5
Stephan-Oterma	1.596	21.4	38.96	1942
Brorsen-Metcalf	0.485	33.0	69.06	1919 Oct 17
Olbers	1.179	32.4	69.57	1887 Oct 8
Pons-Brooks	0.774	33.5	70.88	1884 Jan 26
Halley	0.587	35.0	76.03	1910 Apr 20
Herschel-Rigollet	0.748	56.9	156.0	1939 Aug 9
Grigg-Mellish	0.923	58.6	164.3	1907

Comets exist in many different varieties and with many different orbits. They seem to group roughly into two classes, the short period comets which have orbits with aphelion less than Jupiter's orbit, and long period comets with extremely large aphelion and periods of hundreds of years. It is not possible to determine the orbit of some comets accurately enough to be absolutely sure that the comet is a member of the solar system. Comets are thought to consist of large masses of ice of various sorts, ranging from 0.5 to 5 miles in diameter. As they approach the Sun, solar radiation causes evaporation and creates brilliant gas heads and spectacular tails. The table on the preceding page lists various comets.

No attempt will be made to discuss the many different requirements created by the wide variety of orbits possessed by asteroids and comets. Asteroid velocity requirements will certainly fall within the velocities needed to cover all planetary systems. Likewise, cometary velocity requirements will not be great, providing the orbit is known in advance to sufficient accuracy. By firing a probe to the aphelion of a comet, it would be possible to match trajectories with only small velocity input, and thus to land on or fly formation with the comet during its complete orbit of the Sun.

Newly discovered comets are quite a different matter, as velocity requirements to intercept the comet after detection, but prior to solar passage, could become extremely large.

Synodic Period Effects—One-Way Transfer

Excess velocity capability may be used to open the launch windows to the planets. Mars and Venus present the greater launch window problems, since they have long synodic periods, as already explained. The definition of a launch window at high velocities is more complicated than it might at first seem. A discussion of Mars will be sufficient to illustrate the phenomena involved.

The contours shown in Figure 4-16 are the curves of travel time from Earth to Mars as a function of launch day for a total launch velocity of 60,000 feet per second. The relation-

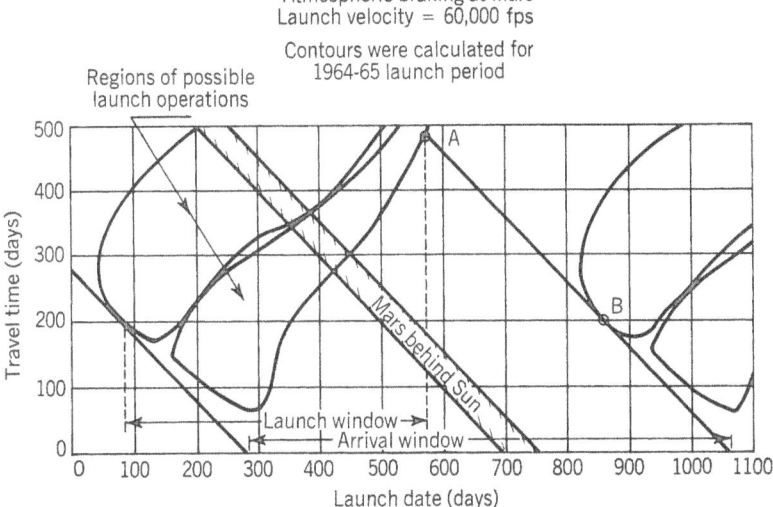

Fig. 4-16.   Earth-Mars launch windows.

ship between two succeeding synodic periods is shown utilizing the same contours as an approximation. Although it is possible to launch at any time of the year with 60,000 feet per second, there is a time (Point A) after which it becomes more sensible to wait until Point B for launch, since the arrival time would be the same. Between Points A and B, one would simply be storing the vehicle in space rather than on Earth. Thus, although a completely open arrival window is available, launching should occur only about half the time.

Additional constraints are evident. A completely open arrival window requires a maximum flight time of 490 days. One might decide arbitrarily to limit this value to some smaller number. If so, both arrival and launch windows will be correspondingly reduced. A plot of launch and arrival windows, as a function of maximum flight time, is shown in Figure 4-17.

At least one other limitation exists. If one uses the completely open arrival window, then part of the time Mars will be on the opposite side of the Sun from Earth. Direct communication is not possible then. Data must be stored for later

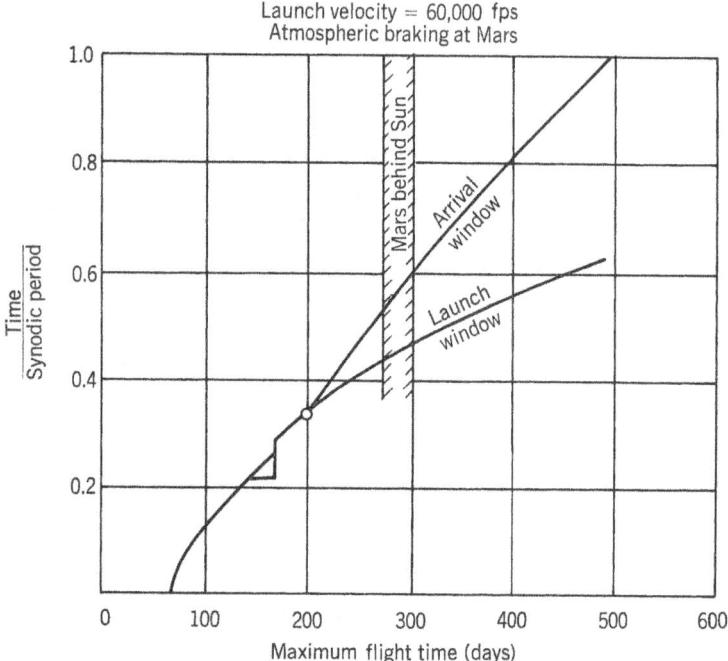

Fig. 4-17.   Earth-Mars launch windows.

transmission, or a communication relay planetoid in solar orbit must be used. Until such a planetoid is available, it is desirable to know when during the synodic period this problem exists.

Accordingly, the band of time during which Mars is hidden from Earth by the Sun is shown in Figure 4-16 and 4-17. Perhaps one should limit maximum flight time to about 280 days. This avoids the problem of Mars being behind the Sun on arrival, and allows arrival windows of approximately 53 per cent of the synodic period and launch windows of 44 per cent. These numbers decrease about 10 per cent if propulsive braking, rather than atmospheric braking, is required at Mars.

A considerable investigation of high launch velocities for

Mars and Venus is required for a number of different synodic periods before these requirements can be ascertained. It appears, however, that 60,000 feet per second launch velocity will permit operation to Mars for at least 40 per cent of the time. This corresponds to about a one-year launch window. Venus should have a more open launch window at comparable launch velocities. Thus, the use of launch velocities as high as 60,000 feet per second can vastly alleviate launch window inconveniences.

Synodic Period Effects—Round Trips

As earlier indicated, one of the very unfortunate effects of the Hohmann Transfer method is the necessity of waiting until the planets are in the proper position for the return flight. The manner in which this is reduced, by using extra launch velocity at both planets, is shown in Figure 4-18. Travel time from Earth to Mars is plotted above the axis with travel time from Mars to Earth below. In both cases, atmospheric braking up to 1.5 V_E is assumed. Several contours are shown up to a

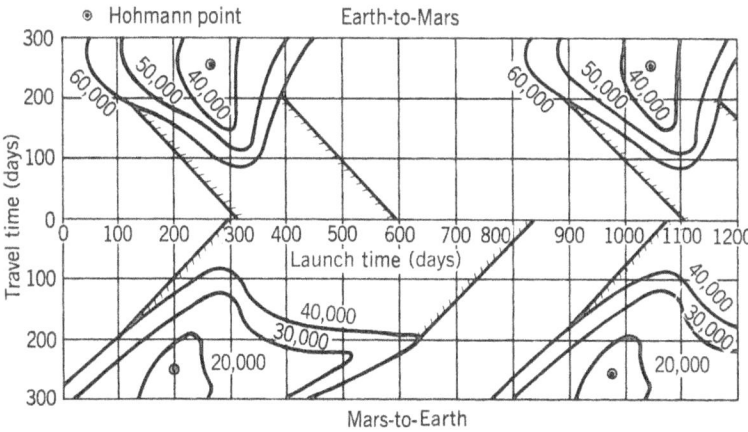

Curve numbers indicate total velocity (fps)

Fig. 4-18.   Round trip synodic period effects.

maximum of 60,000 feet per second launch velocity for Earth and 40,000 feet per second for Mars. Typical operational regions for maximum travel time of 200 days are shown. With 100,000 feet per second total velocity, operation from Earth to Mars would be possible 36 per cent of the time and from Mars to Earth 82 per cent of the time. Regardless of the velocity, travel during favorable periods can be done with shorter flight times or greater cargo loads. Even among the planets, there will be tourist seasons.

Curves of the sort shown in Figure 4-18 must be calculated to and from all planets of interest for various launch velocities to ascertain the exact velocities required for interplanetary flight. One must consider the use of swingbys of other planets, and some cases where mid-course impulses—either nearer or farther from the Sun—reduce the velocities. It is not possible to present the velocity requirements for interplanetary flight in a few simple curves. Voluminous calculations of many actual cases have now been made and the interested student can find many sets of curves similar to those seen in Figure 4-18.

Lunar Refueling

If it were possible to refuel rocket vehicles on the moon, large reductions in total vehicle weight required for transporting cargo to and from the moon would be possible. The total impulsive velocity requirement to go to the moon and return with atmospheric braking at earth is about 58,000 feet per second. Of this, 30,000 feet per second is required to reach earth orbital velocity, 11,400 feet per second to inject from earth orbital velocity to lunar transit velocity and provide for guidance corrections, and 18,600 feet per second for landing on the moon and returning. Instead of one 58,000 foot per second rocket vehicle, imagine using two vehicles performing the same mission—one earth-based and one moon-based. One could use a 30,000 foot per second vehicle to take the payload to earth orbit, have a 28,000 foot per second lunar-based vehicle come to earth orbit (using aerodynamic braking to enter earth orbit), pick up the payload, return, and land on the moon. Thus, two vehicles of roughly half velocity capability would be used to do the job of one vehicle.

The effect of doing this is startling. A single stage vehicle using high energy chemical propulsion would require a gross weight to payload ratio of about 20 to reach 30,000 feet per second. It would require very light structures to do this without staging. If completely earth-based vehicles of this sort were used, a total gross weight to payload ratio of 20 × 20 or 400 would be required. If a lunar-based vehicle picked up the payload from the earth-based vehicle, 20 + 20, or 40, would be the total gross weight to payload ratio required. The total rocket weight to perform the operation would be reduced by a factor of 10. This strong effect occurs because fuel already at a place (the moon), which requires a large amount of energy to reach from earth, is essentially fuel which already has that much additional energy content with respect to earth. To an observer on earth, the lunar-based rocket looks like a very high energy vehicle.

It is frequently stated that the establishment of large space stations on earth orbit is the key to planetary exploration, since special vehicles can be prepared and launched from these stations. Compared to lunar-based vehicles, this is a highly debatable point, simply because there is not a fuel supply on earth orbit. If we compare the two cases after each base is established—a space station for launching from earth orbit and a lunar refueling and launch base—the expenditure of 8300 feet per second by the lunar-based vehicle results in the generation of earth escape velocity and no fuel need be brought from earth to accomplish this. For the orbital-based rocket, the generation of 10,400 feet per second is required to create earth escape velocity and the fuel with which to do this as well as the remainder of the mission must be ferried up from earth.

It is more difficult to get the space vehicle to the moon in the first place than it is to place it in the earth orbital facility. Since lunar-based refuelers can help it get to the moon, however, this initial fuel expenditure is only a factor of two greater than that for the orbital case.

The initial creation of the lunar facility would be expected to be substantially more difficult than the earth orbital facility, since much higher total velocities must be used prior to the time that refueling capability exists on the moon's surface. The establishment of a refueling facility might be done relatively quickly, with luck, or it might require years or decades of scientific research and engineering prospecting before the proper materials could be found. Therefore, the relative initial expenditure involved is highly speculative.

In both cases, the continuing expenditure of supplies necessary to operate the facilities can have a strong effect on their utility. Food, air, and equipment necessary to maintain the facility must all be brought from earth. If the amount of supplies required per pound of fuel available becomes too high, then the utility of either system, compared to direct earth launch, will be seriously eroded. In the case of the lunar facility, it will require twice the fuel expenditure to bring supplies to the moon, even if lunar-based rockets are used in the process. This disadvantage would be countered to the extent that success was achieved in obtaining food, water, or oxygen from the moon itself. The point is simply that the case for the earth orbital station is not obvious. Under some circumstances, a lunar base would have greater utility.

The possible use of lunar refueling is a fascinating prospect. In the exploration of this planet 500 years ago, living off the land was a common technique of explorers. It was important to the crews to find food and fresh water for re-supply at their destination. In space exploration, the supplying of food, water, and air for the crew may not be nearly as important as the care and feeding of the rockets. The importance of finding fuel supplies on the moon which can be used in supporting whatever transportation system is utilized for travel on the moon itself is obvious. In light of modern technology, transporting fuel from one planetary body to another for any great length of time in order to supply local transportation needs on the second body would seem ridiculous.

Planetary Refueling

Virtually everything which has been said about a lunar refueling base applies to all the planets. In the minor planets—Mars, Venus, Mercury, and Pluto—such bases would have to be on their surfaces, except for a possible base on one of Mars' two small moons. It remains to be seen whether Pluto's surface is too cold or Venus' too hot for such operations. Bases on the major planet's forbidding surfaces appear out of the question, but fortunately, they all have natural satellite systems around them. All, except Uranus, have at least one natural satellite of the same size category as our Moon. It would, therefore, seem logical to have refueling bases on these satellites. Were such bases established, then at some future date it would be easier to penetrate to the planet's surface, with the operations mounted from the refueling bases on the satellites.

As an example of the comparative benefits to be gained by refueling at a planet, we can examine the case of Mars. The Earth launch velocity contours (shown in Figure 4-18) are reduced by 28,400 feet per second if launch occurs from the Moon. A rocket of 30,000 foot per second velocity capability operating between the Moon and Mars, which could be refueled at either place, is equivalent to an 88,400 foot-per-second non-refueled Earth-based rocket. It could maintain operation during about 33 per cent of the synodic period, with maximum travel time of about six months. A far better operation than is commonly assumed could be mounted between Earth, Moon, and Mars, using chemical rockets, if it were possible to refuel at each place. It would seem that with such great gains, the establishment of refueling bases of high efficiency should be one of the highest priority aspects of any space program.

Liquid Propellant Rockets

Liquid rockets suitable for space applications have already started to take on a different complexion from those developed for ballistic missile use. The velocity region is higher; therefore, the need for high specific impulse is greater. Though more exotic propellants may be harder to handle, their use is often justified. Space missions do not always require the instant alert capabilities of weapons, and it is possible to put up with less convenient propellants if their performance is higher.

The cargo to be carried to space is much greater than that carried by ICBM's. Nuclear warhead technology can make small ICBM's equivalent to large ones, but man cannot be miniaturized, and the paraphernalia required for his work and pleasure, no matter how greatly reduced, is endless. For fundamental reasons, space rockets will be larger and of higher performance than terrestrial weapon delivery rockets.

Space Storable Liquid Propellants

Storage in the vacuum of space, when subjected to solar radiation and micrometeorite hazards, is quite different from storage at the bottom of the Earth's atmosphere. The extremes of solar radiation flux presented to propellant tanks will be far greater than is achievable on Earth. Eventually we will want propellants which are storable as far in toward the Sun as Mercury is and as far out as Pluto. The stabilized temperatures achieved by a properly shielded space container may be so cold that a cryogenic propellant refrigerated on Earth is suitable for storage in space.

A number of propellant combinations have been suggested as space storable liquid propellants which need no special insulation. Some are listed in the table on page 120. The performance of some of these combinations is quite high. Oxygen difluoride-diborane has a vacuum specific impulse over 400 seconds, which compares reasonably well with high energy cryogenic propellants. Only a small amount of actual engine development work utilizing any of these propellant combinations has occurred. There is, however, an obvious need for some form of space-storable propellant in the future.

Any solid propellant competing for these applications must produce a vacuum specific impulse in the vicinity of 400 seconds, and be able to stand the temperature environment likely to be encountered after extended soak periods in deep space.

Since low temperature has always been one of the problems of solid propellants, new ideas in the solid propellant field, comparable to the introduction of modern chemicals and case bonding grain techniques, will be required if solid propellants are to compete with space storables.

Cryogenic Liquid Propellants

The fact that the United States was not first into space was bad enough. When it became clear that a rocket much larger than anything that existed in the United States was being used at that time, it was an additional blow to the American ego. Whether this larger Russian rocket was a good or bad weapon was beside the point. It was still a good space rocket. Consequently, the United States plunged into the development of larger rockets with traditional vigor, a vigor so great that people now seem to equate rocket performance merely with size, regardless of the velocity achieved.

One of the rocket engines, put into development as a possible component for future large rockets, was the F-1 engine. It produces 1.5 million pounds of thrust using liquid oxygen and kerosene as propellants. The engine by itself is 40 per cent as long as and over 50 per cent as heavy as a loaded V-2. The nozzle exit diameter is double a V-2's maximum diameter. The turbine driving the turbo-pump produces about 50,000 horsepower. As of 1965, a number of these monsters had been successfully run in static tests for durations of three minutes. The F-1 is, as far as is known, the largest liquid rocket engine in existence. Five of them will power the first stage of the Saturn V moon rocket which will weigh six million pounds at take-off. The exhaust power of the five engines will be about 57 million horsepower at launch. The five-engine cluster has been successfully tested.

High Energy Cryogenic Propellants

In spite of the need for large rocket engines, it is clear that higher performance engines are also required. That hydrogen was one of the most energetic of rocket fuels had been known for a long time. Enough experimental work had been done on the combination of hydrogen and oxygen by the time of Sputnik to indicate the feasibility of placing this propellant combination into service. The advantages are many. This combination seems to burn smoothly with a minimum of combustion instabilities. The specific impulse is substantial compared to oxygen-kerosene. Hydrogen-oxygen in vacuum gives about 430 seconds specific impulse (see table on page 120).

As usual, there are disadvantages. Hydrogen and oxygen form an explosive mixture over an unusually wide range of mixture ratios and fear of explosive hazard was high. Of all the cryogenic materials, hydrogen has the lowest boiling temperature except for helium. It boils at only 37° F. above absolute zero which is well over 100° F. colder than liquid oxygen. The handling requirements and insulation techniques would be expected to be more severe with hydrogen than with oxygen.

Hydrogen is the lightest substance known. While the oxygen-kerosene combination has roughly the same density as water (about 63 pounds per cubic foot), liquid hydrogen weighs only four pounds per cubic foot. This penalty of a factor of 15 in tank volume is alleviated by the fact that hydrogen-oxygen engines utilize about five times as much oxygen as hydrogen. The total tankage for a hydrogen-oxygen rocket requires about three times the volume of a comparable weight of oxygen-kerosene.

In spite of the volume disadvantages and explosive hazards involved, the performance gain seemed worth the effort of putting hydrogen-oxygen rockets into the inventory. It has been borne out to be a proper decision. Hydrogen requires special handling, but so do oxygen, nitric acid and other rocket propellants which have been successfully utilized. Insulation techniques at very low temperatures have been forthcoming. Explosions have been rare. Utilization of hydrogen in rockets does not cause any greater difficulties than other propellant combinations.

The first hydrogen-oxygen engine of appreciable size was the RL-10 engine with a vacuum thrust of about 15,000 pounds. Two of these power the Centaur and six of them the

Saturn S-IV upper stages. Both have been flown successfully to orbit. Under development is another hydrogen-oxygen engine, the J-2, with a thrust of around 200,000 pounds. This engine will power the Saturn S-IVB upper stage, and five of them will power the Saturn S-II second stage. A hydrogen-oxygen engine with a thrust of about 1.5 million pounds, the M-1, has been partially developed as a possible engine for future vehicles.

Clearly, hydrogen-oxygen will be used in the immediate future. This brings up the question of whether or not better high energy liquid chemical propellants might be available. The table on page 120 shows some possibilities. Hydrogen-fluorine has the highest specific impulse of two propellant combinations, about 20 seconds higher than hydrogen-oxygen. This does not seem a very spectacular performance improvement. Hydrogen-fluorine, however, burns at very high mixture ratios. It is possible that 12 pounds of fluorine per pound of hydrogen or more will be achieved. The amount of hydrogen carried would be small enough so that the large volume penalties of hydrogen tanks in a hydrogen-oxygen rocket would be alleviated. Therefore, the λ's of a hydrogen-fluorine stage could be higher than a comparable hydrogen-oxygen stage.

To date, high-energy propellants have been used only as upper stages, but it is worth considering their use in first stages as well. Though many people view this as inappropriate, there are no valid reasons against it. Using hydrogen-oxygen in the first stage will yield not a smaller vehicle, due to the required large size of hydrogen tanks, but a substantially lighter one. The use of hydrogen-fluorine in all stages of a vehicle will yield both a lighter and smaller vehicle. For earth orbital velocities, the difference between hydrogen-fluorine and hydrogen-oxygen is not great. For lunar mission or for solar escape velocities, however, the effects can be large.

Rockets sketched in Figure 4-19 represent different designs to meet lunar landing and return requirements. Two rockets are shown compared to the Apollo/Saturn V. One uses hydrogen-oxygen and the other hydrogen-fluorine in all stages,

Theoretical Liquid Propellant Performance
Equilibrium Flow (Frozen Flow)

OXIDIZER	FUEL	VACUUM, $\epsilon = 40$			SEA LEVEL
		Mixture Ratio	Specific Gravity	I_{sp} (seconds)	I_{sp} (seconds)
OXYGEN*	Hydrogen*	4.50	0.31	456 (446)	391
	Hydrogen*-Beryllium	0.92	0.24	536 (526)	458
FLUORINE*	Hydrogen*	9.00	0.50	475 (450)	411
	Hydrogen*-Lithium	0.85	0.17	505 (479)	432
	Hydrazine	2.34	1.31	424 (369)	364
	Ammonia	3.31	1.12	416 (365)	360
OXYGEN-DIFLUORIDE**	Hydrogen*	7.00	0.43	465 (446)	401
	Kerosene	3.80	1.28	396 (348)	341
	Hydrazine	1.62	1.27	397 (354)	339
	Diborane**	3.82	1.00	429 (380)	365
NITROGEN-TETROXIDE	Hydrazine-Beryllium	0.45	1.18	384 (372)	326
	MMH-Beryllium	1.08	1.20	376 (355)	315
	MMH-Aluminum	0.83	1.30	351 (397)	339
HYDRAZINE	Diborane**	1.16	0.63	401 (397)	339
	Pentaborane	1.26	0.79	390 (381)	328

* Cryogenic
** Space Storable

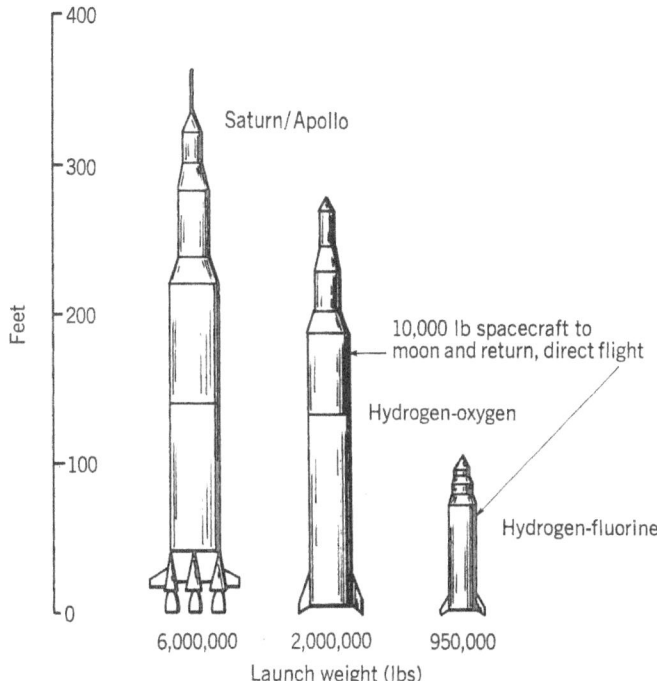

Fig. 4-19.   High-performance chemical rockets.

including those which land and take off from the moon's surface. Both are able to take a 10,000-pound payload (heavier than the Gemini capsule) to the moon and back by direct flight. λ's are assumed roughly appropriate for each propellant combination. The hydrogen-fluorine rocket could do the lunar mission with less than one million pounds launch weight, or for less than the size of the Saturn I vehicle instead of the Saturn V.

There are many reasons why the United States did not pursue an all hydrogen-fluorine moon rocket. The state of fluorine engines was nowhere near as advanced as hydrogen-oxygen. The largest hydrogen-fluorine thrust chamber fired as of 1965 is a 40,000 pound thrust unit.

In using either hydrogen-oxygen or hydrogen-fluorine engines for lunar landing and take-off, fears have been expressed about the inability to insulate successfully the cryogenic propellants for a sufficiently long period of time. The development in recent years of super insulations, however, indicates that insulation problems could be solved. Super insulations work by using multiple layers of foils as radiation barriers, usually with a vacuum of space between foil layers to reduce convection and conduction. When landing during the lunar night, there will not be a large thermal input from the lunar surface. If landing occurred during the lunar day, some sort of sunshade could be erected.

Fear of using fluorine in the first stages is based on its toxicity. It could be dangerous if released in quantity in case of an accident. One of the principal products of the combustion of hydrogen and fluorine is hydrofluoric acid which is also a toxic compound. Fluorine, however, is no worse than a number of propellants used in the past. Nitric acid is, after all, nothing to be eaten for dinner; yet, many anti-aircraft missiles using this propellant were at one time deployed in the United States.

The comparison between fluorine and nitrogen tetroxide is particularly interesting. The concentration which causes damage is essentially the same. Fluorine's odor is readily detected when its concentration reaches the level which is toxic. Nitrogen tetroxide is virtually odorless. Thus, it would seem hard to believe that fluorine is actually as dangerous as nitrogen tetroxide to use around people. Yet we do not seem concerned about using nitrogen tetroxide as one of the propellants in the Titan II launch vehicle which has been used to launch the two-manned Gemini capsule to orbit.

It is unfortunate that fluorine etches glass and that almost every technical person has learned this while studying high school chemistry. Every one knows fluorine is toxic. It is compatible with many different metals and vast quantities have been manufactured and used without any trouble. Some day we may see manned lunar operations performed with the relatively small hydrogen-fluorine rockets shown in Figure 4-19.

Exotic Liquid Propellants

Hydrogen-fluorine rockets are by no means the end of the chemical rocket development stream. Several interesting three-chemical combinations, known as tripropellants, are being investigated. (Some of these tripropellants are shown in the table on page 120.) Theoretical calculations indicate that hydrogen-oxygen-beryllium should produce a specific impulse over 500 seconds. Up to the present time, only small combustion tests have been made with tripropellants, and the degree to which theoretical performance can be realized is as yet uncertain.

It is conceivable that hybrid rockets will be a way of utilizing tripropellants. In spite of their low molecular weights, both beryllium hydride and lithium hydride are solid materials, as is beryllium oxide. The previous comments that solid rockets probably cannot achieve very high energy performance applies because both fuel and oxidizer cannot be found in solid form. A portion of it can, so hybrid rockets may be able to match liquids. Which type will achieve the highest combustion and nozzle efficiency is only a matter of speculation.

Other combinations are of interest. Ozone gives about eight per cent more specific impulse than oxygen with either kerosene or hydrogen. No one has succeeded in stabilizing ozone to a degree which seems safe to handle. Some research has been done on the use of both free radicals and metastable compounds as even more energetic chemical propellants. Use of the highest energy-free radical known—atomic hydrogen—would yield a specific impulse of about 1200 seconds. As with ozone, no convincing progress in handling these highly reactive systems has yet been reported.

Undoubtedly, under the impetus of the thrust into space, much more will be learned about these combinations in the next few years. If rocket development history is any indication, means will be found to use some of the exotic combinations, and a further increase in specific impulse of high energy chemical propellants can be expected.

New Types of Engines

The basic shape of rocket engines has changed little since Goddard's early experiments. Prior to that, rockets were very low-pressure devices which frequently used paper cases and had no effective nozzle. Goddard contributed the high-pressure case and added a DeLaval nozzle. Ever since, rockets have had cylindrical or spherical chambers with DeLaval nozzles at the back. Many improvements in nozzle design have been made, and modern bell-shaped nozzles are more efficient and of shorter length than the original conical nozzles.

The chamber pressure of rocket engines is a source of constant discussion among rocket propulsion engineers. With solid engines, all propellant is contained within the thrust chamber, and it is not possible to change appreciably this chamber volume. It is, therefore, usually desirable to develop grains which burn efficiently at low pressures in order to reduce total engine plus propellant tank weight. It would seem natural that the same trend would follow in liquid engines. Lower combustion pressures would result in lighter weight thrust chambers, and pump power and weight would be smaller if a lower pressure were used (see Equation 1-16).

Later engine designs dispute this logic. If a higher chamber pressure is used in a liquid thrust chamber, the size of the chamber is reduced. The throat area comes down by inverse proportion to the chamber pressure and other dimensions come down accordingly. The stress in the wall of a spherical container is given by the expression:

$$\sigma = \frac{pr}{2t} \qquad (4\text{-}7)$$

where σ = material stress in pounds per square inch; p = internal pressure in pounds per square inch; r = radius of sphere in inches; and t = wall thickness in inches. This expression also applies to the longitudinal stress in a cylinder, although the hoop stress is twice the value. For all shapes, the stress is proportional to pr/t. The weight of pressure shell will be simply its area times thickness, or making use of Equation 4-7:

$$w_{co} \sim A_{co}t \sim \frac{A_{co}pr}{2\sigma} \qquad (4\text{-}8)$$

where w_{co} = engine chamber weight; and A_{co} = surface area of chamber. For similar shaped engines of constant thrust, A_{co} times p is constant just as throat area times p is constant. For equally stressed chambers the weight is decreased proportionally to the radius. The saving in total surface area which must be covered more than makes up for the additional material thickness to contain the higher pressures.

The actual situation is more complicated, but Equation 4-8 indicates how one can obtain a lighter thrust chamber. Weight penalties in modern turbo-pumps are sufficiently small that very high pressure engines plus turbo-pumps are still considerably lighter than their lower-pressure equivalents. This permits an additional improvement in specific impulse because some of the space and weight decrease can be used to include a higher expansion ratio nozzle, which will give a higher specific impulse in vacuum. With high internal pressure, the nozzle is more properly expanded at sea level even with a high expansion ratio. A typical normal engine has $p_c/p_{at} = 20$ at sea level while a high pressure engine might have 200. A high pressure engine, consequently, does not have as great a variation of specific impulse with altitude as a low-pressure engine (see Figure 1-4). Serious proposals exist for engines with operating pressures of 3000 pounds per square inch, or about 10 times the operating pressure of current liquid rocket engines. This is one promising approach to new engine designs.

High pressure engines with conventional nozzles still look like a relative of Goddard's rocket engines. Other types of engines being investigated do not remotely resemble rocket engines. Many attempts to cut nozzle length drastically, or to use the aft part of the rocket vehicle as part of the nozzle, have been considered. Some possibilities are shown in Figure 4-20. The expansion deflection nozzle works by ejecting the hot gases sideways where they are then turned by a suitably curved surface in a shock expansion process. It looks like a short, conventional rocket.

The family of plug nozzle engines are inside-out rocket engines, with the combustion chamber a large ring around the base of the vehicle and the flow expanding into the base behind the vehicle. Pressure forces on the base correspond to the pressure forces inside the expansion cone of a normal nozzle. The plug-nozzle concept can be made to function with a continuously toroidal combustion chamber, with linear segmented chambers, or by putting individual high pressure engines around the base. In the aerodynamic nozzle, the plug is created aerodynamically by injecting about two per cent of the flow into the center of the base at low pressure. The effect of the interaction of the two flow patterns is almost as if there were a physical plug in the center.

Plug nozzles of all types are altitude-compensated to a large extent. They behave almost as if an optimally expanded nozzle were used at each altitude. This occurs, as illustrated in Figure 4-20, because the exhaust flow is free to form only a narrow tube at high pressures, but equally free to expand outward at low pressures.

–·— Low altitude exhaust
––– High altitude exhaust

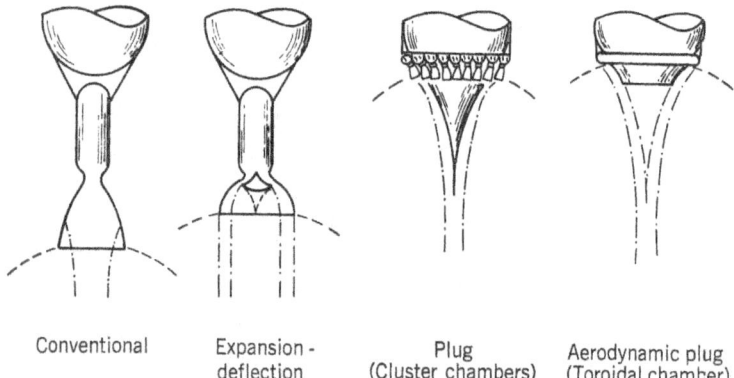

| Conventional | Expansion -
deflection | Plug
(Cluster chambers) | Aerodynamic plug
(Toroidal chamber) |

Fig. 4-20.   New engine types.

In the case of a toroidal engine with aerodynamic nozzle, the complete engine plus nozzle looks like a small spare tire wrapped around the end of the fuel tank. They may not look like rocket engines but there is no doubt that recent ideas on building rocket engines have exciting possibilities in terms of better specific impulse, lighter weight, and greater convenience in overall vehicle design. It would not be surprising if most rocket engines a decade hence were to look vastly different from the standard rocket engine of today.

Nuclear Thermal Rockets

The conversion of mass to energy is governed by Einstein's famous equation:

$$E = mc^2 \qquad (4\text{-}9)$$

where c = velocity of light in feet per second. Putting the velocity of light (983 million feet per second) into Equation 4-9 gives 3×10^{16} foot-pounds as the energy which would be released by the complete conversion of a pound of mass into energy. This is 17 billion times the energy release per pound of smokeless powder.

Only a portion of this energy is released in nuclear reactions. In fission reactions, large nuclei such as uranium 235 are split apart by neutrons. In the process, other neutrons are released which in turn split other atoms. The weight of all the pieces is less than the original atoms by the amount of energy produced. In fission reactions, from 0.072 to 0.12 per cent of the mass is converted to energy. This, if it could be converted to exhaust velocity with perfect efficiency, would give a specific impulse of from 1.16 to 1.50 million seconds (see Equation 1-12). If the amount of nuclear material is too small, large surface area compared to mass will allow too many neutrons to escape, and a sustained chain reaction will not be possible. The minimum mass required for nuclear fission is called the critical mass.

In fusion reactions, light nuclei such as hydrogen are forced together by high temperature impacts and fuse into heavier nuclei such as helium. The weight of the final helium is less than the original hydrogen by the amount of energy released. Fusion reactions typically convert from 0.7 to 0.9 per cent of the mass to energy, or about 10 times the energy per pound of fission reactions. The specific impulse with perfect efficiency would be from 3.6 to 4.1 million seconds. As of 1965, sustained controlled fusion reactions have not been achieved.

The fission process, unfortunately, is not as simple as described. When atoms fission, they may break in several different ways. Hence, a variety of pieces, fission products, result. Some are radioactive and must be safely handled. In addition, such nuclear radiations as beta particles (high velocity electrons) and gamma rays (high energy x-rays) are produced. Even if fission products were no problem, radiation and stray neutrons would heat structures and propellants, damage equipment, and induce radioactivity in the structures unless selected materials were used. Use of nuclear reactions involves a shielding penalty not suffered by chemical rockets. Despite shielding problems, the tremendous energy release of nuclear rockets makes their eventual utilization almost inevitable.

The most obvious way of using nuclear energy in rockets is to put a small reactor in the rocket chamber and use it to heat a propellant. This is known as a solid core nuclear rocket. Figure 4-21 shows such a rocket concept. The first nuclear rocket, actually a test device too heavy to fly and hence named Kiwi, was fired in Nevada, USA, on July 1, 1959.

Unfortunately, a solid core nuclear rocket is limited in its performance potential. To obtain a high exhaust velocity in a thermal rocket, it is necessary to heat the gas in the combustion chamber to a high temperature (see Equation 1-14). The nuclear reactor must be hotter than the propellant; otherwise, heat will not flow to it from the reactor. The highest temperature which can be reached within a solid core rocket is the temperature at which the reactor would melt. Thus, materials considerations limit the temperature which can be achieved. The highest known melting point of a material (hafnium carbide) is about 7500 degrees F.

For a given temperature, the highest specific impulse is

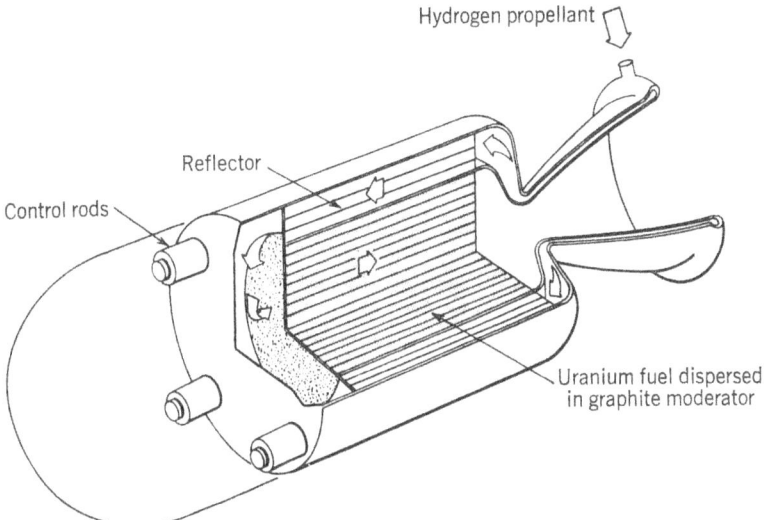

Fig. 4-21.   Graphite solid-core engine. (Douglas Aircraft Co.)

obtained with the lightest molecular weight propellant (see Equation 1-14). The most logical propellant to use is hydrogen—the lightest known substance. Heated to about 5000° F, hydrogen gives a specific impulse of about 1000 seconds. For this specific impulse, Equation 1-11 gives 16.1 million foot-pounds as the kinetic energy of the exhaust per pound of propellant. Hence, the materials limitation on a solid core nuclear rocket is so severe that less than one-millionth of the fission energy release per pound possible can be placed in the exhaust jet.

Moderated Solid Core Rockets

Operating reactor temperatures must be substantially below material melting temperatures. The material must contain nuclear fuel, and this will usually adversely affect the material. The reactor must retain adequate strength, and materials become weaker at higher temperatures before melting. Materials which can be used also are limited by their nuclear characteristics. The neutrons which sustain the reaction are affected in different ways by different materials. Some materials absorb neutrons and act as poisons to the reaction. Others reflect neutrons. Others, such as graphite, are known as neutron moderators.

Not all neutrons produced by a fissioning process are alike. Some are produced immediately and called prompt neutrons. Others are delayed, appearing as much as minutes later. Neutrons are given off with a wide spectrum of velocities. The fast ones are called fast spectrum or fast neutrons. All prompt neutrons have relatively high velocities and are classified as "fast" neutrons. If the neutron's velocity is such that its temperature is the same as the material through which it is moving, the neutron is said to be in thermal equilibrium with its surroundings. It is called a thermal neutron. Thermal neutron velocities are only about 0.05 per cent of the velocity of the slowest "fast" neutrons.

Thermal neutrons react more readily than fast neutrons with the uranium they pass through. A reactor which operates with thermal neutrons can have a smaller critical mass than a reactor using fast neutrons. A moderator is a substance which slows down (moderates) the velocity of the neutrons. A thermal reactor carries much less uranium for critical mass, but must pay the penalty of carrying the moderator. Also it cannot contain any elements which are poisonous to thermal neutrons. Since many metals are poisonous to thermal neutrons, the number of them which may be used in thermal reactors is seriously limited.

Nuclear rockets under active development in the United States as of 1965 (Nerva) are of the graphite-moderated thermal variety. The thrust is about 55,000 pounds, the specific impulse about 745 seconds, and the power of the reactor about 1100 megawatts. Although 1100 megawatts seems small (producing only 55,000 pounds of thrust), it is 1.1 million kilowatts (1.5 million horsepower), or enough power to supply the average needs of a city of 500,000 people if produced continuously and converted to electricity with perfect efficiency.

A typical, large stationary nuclear power plant now operating—the Dresden Station in Morris, Illinois, USA—produces only 700,000 kilowatts of thermal energy (208,000 kilowatts of electricity). Although it can run continuously (rocket reactors cannot), the volume of the Dresden reactor core is much larger than the volume of the rocket cores. For suitable rocket propulsion, reactors must be so compact they operate at power densities well over 10 times those of previous reactors.

Solid core reactors other than graphite have been suggested. One possibility is to build thermal reactors which are metallic. Since the metal used must be of extremely high temperature capability, only a few are suitable. Tungsten is one since its melting point is about 6090° F. Normal tungsten is quite poisonous to thermal neutrons; however, tungsten-184 has a low absorption cross section to thermal neutrons, and would be suitable for thermal reactor construction. About 30 per cent of natural tungsten consists of tungsten-184. Its occurrence is not as rare as uranium-235 is in the more common uranium-238 (less than one per cent), but tungsten-184 still must be separated from normal tungsten by the same procedures required for uranium-235. Consequently, tungsten-184 will be quite expensive if used in reactors.

Solid core reactors have also been suggested which do not have the fuel embedded in the moderator. These are known as heterogeneous reactors. Since energy is not released in the moderator, low temperature moderator materials, such as water, may be used, while high temperature materials, like tungsten, contain the hot fuel. Such reactors may be of interest although obviously more complicated than homogenous reactors.

Fast Solid Core Rockets

A metallic reactor which operates on fast neutrons could be considered. The inventory of uranium which must be carried will be higher; however, the lack of a moderator results in smaller reactors. Normal tungsten and many other metals are possible materials for construction of such reactors.

Among the possible advantages for fast metallic reactors is their reusability, which is greater than that of thermal reactors. The reason is that one of the fission products produced in uranium fissioning—xenon-135—is poison to thermal neutrons. The xenon-135 produced shortly after shutdown may prevent restarting of a thermal reactor for many hours. Since xenon-135 is not a poison to fast neutrons, a fast reactor would not have this problem. Large reactors, however, tend to be thermal. Reactor design is limited by the ability to remove heat from the core. If one builds too large a fast reactor, it becomes impossible to cool. It would then be necessary to add poisons to the reactor. One might as well add moderator instead and save valuable nuclear fuel.

Difficulty of control is often erroneously quoted as a problem of fast reactors. The reason for this is a confusion between fast-spectrum, i.e., high-velocity neutrons, and those which are produced immediately (in fast time)—prompt neutrons. Reactor control is frequently simplified by making use of the delayed neutrons. If the reactor is designed so that delayed neutrons are required for criticality, their population can be adjusted by a relatively slow reacting control system. The proportion of delayed neutrons is obviously the same in fast and thermal reactors, however, and the dynamics of the control problem with both types of reactors are essentially identical. Fast reactors are in service for stationary power plants and submarine power systems. For some reason, they are more common in Europe than in the United States.

Radioisotope Heated Solid Core Rockets

Another possibility for building solid core nuclear rockets is to heat the propellant by means of radioisotopes. Isotope heating has been used to supply electrical power. It can be used likewise to heat propellants. As in other solid core rocket systems, the specific impulse is limited to 800 or 900 seconds.

Several radioisotopes are quite powerful, as shown in the following table. Polonium-210 produces 134 watts per gram, or 61 kilowatts per pound of isotope fuel compound. To produce a pound of thrust with a specific impulse of 1000 seconds requires 21.8 kilowatts (see Equation 1-11). The polonium alone weighs 36 per cent of the thrust, and current thrust units are about 300 times the weight of the isotope. With the best isotope shown, the thrust/weight ratio would be less than 0.01. The effect of such low thrust/weight ratios will be examined in the next chapter.

Isotopic Heat Sources

Parent Isotope	Half-Life (years)	Type of Decay	Specific Power (Watts/gram)		Shielding
			Pure	Fuel Compound	
Strontium-90	28	Beta	0.90	0.38	Heavy
Cesium-137	30	Beta/Gamma	0.42	0.067	Heavy
Promethium-147	2.7	Beta	0.33	0.27	Minor
Plutonium-238	89	Alpha	0.56	0.39	Minor
Curium-244	18	Alpha	2.8	2.49	Moderate
Curium-242	0.45	Alpha	120	98	Minor
Polonium-210	0.38	Alpha	141	134	Minor
Cerium-144	0.78	Beta/Gamma	25.6	3.8	Heavy
Cobalt-60	5.3	Beta/Gamma	17.4	1.7	Heavy

Only recently have radioisotope heated rockets received attention. In spite of low thrust/weight ratio, availability and cost problems, they may have utility in certain missions, perhaps in high performance unmanned probe vehicles. The alpha particle (helium nuclei) emitters are of particular interest since they require almost no shielding. Alpha particles cannot even penetrate clothing. Alpha emitters exist with almost any half-life desired down to less than a second. Since power density increases as half-life decreases, very powerful alpha emitters exist. Sufficiently rapid production and incorporation into rockets would be difficult, even if the launching site were next to the isotope factory. The ability of a reactor to be started or stopped at will is invaluable, and the use of radioisotopes for main drive rockets in large vehicles does not seem appropriate.

Shielding

One might intuitively expect the shielding weight of nuclear rockets to be so high as to cripple their utility. Shielding weight was a major problem of the nuclear aircraft program and rocket reactors must be more powerful, while rockets are notoriously more sensitive to weight than airplanes. This viewpoint seems rational, but such is not the case. Figure 4-22 illustrates the pertinent facts involved.

Although the reactor power of an aircraft would be several hundred times lower than that of a rocket, shielding weight is

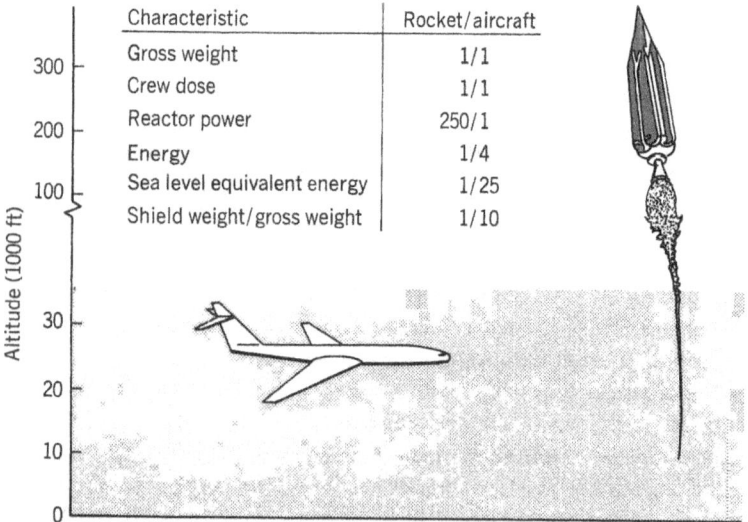

Characteristic	Rocket/aircraft
Gross weight	1/1
Crew dose	1/1
Reactor power	250/1
Energy	1/4
Sea level equivalent energy	1/25
Shield weight/gross weight	1/10

Altitude (1000 ft): 300, 200, 100, 30, 20, 10, 0

Fig. 4-22. Nuclear vehicle shielding comparison. (Douglas Aircraft Co.)

a function of energy rather than power. The purpose in building nuclear aircraft is to obtain long duration flights so the operating time is days or weeks. If an airplane similar to the jet transport discussed in Chapter 2 were driven with nuclear engines for a week, the velocity equivalent of the energy consumed (see Equation 2-8) would be about 50,000 feet per second, or twice escape energy. Since rockets are twice as efficient (Chapter 1), the nuclear airplane consumes four times the energy of an escape velocity rocket. Furthermore, the airplane by definition must always operate in the earth's atmosphere, hence is continually subject to the radiation scattered back from the earth's atmosphere. This scattered radiation accounts for the largest contribution to the shielding weight. The rocket, on the other hand, climbs quickly out of the atmosphere. Estimates of the sea-level equivalent energy that each device would produce (a measure of the total scattered radiation) indicate a factor of 25 in favor of the rocket.

Further reductions in shielding weight can be made by utilizing cargo, food, the life-support system, and propellant in the tanks as shielding material. The nuclear rocket shield system is also useful as a space radiation shield, an item which it may be necessary to add to chemical rockets for some missions.

Rocket Orbital Transports

New engines always raise the question of the desirability of new types of vehicles. One possibility is a reusable earth orbital transport. We would have had single stage rockets of orbital velocity long ago if they were easy to build. Even with high energy chemical propellants, the structural weights required have seemed too low to produce with confidence. With an altitude-compensated engine and high-energy chemical propellants, the goal of single stage rockets to orbit appears nearer.

Many years ago, the Atlas ICBM design was conceived as simpler than a two-stage missile. The idea was to have a single propellant tank but to discard several engines at staging. The reasoning was that the tank weight could be made extremely light with pressure-stabilized structures, but the engines were heavy. By discarding only engines, most of the staging effect could be obtained. This was labelled a one-and-a-half stage missile.

Should high pressure engines have a very high thrust/weight ratio, little would be gained by discarding the engines. Altitude-compensated nozzles obviate the previous need for special low altitude and high altitude engines. Reusable structures would be expected to be heavier than normal rocket tanks, since they must stand normal rocket loads as well as atmospheric temperatures and re-entry loads. A logical design may have a relatively heavy, reusable structure with altitude-compensated engines but with a significant portion of fuel in auxiliary tanks which are jettisoned part way to orbit. It would not be surprising if new engine designs stimulated this type of vehicle.

The propellant weight required for an orbital transport (and lunar transport) is shown in Figure 4-23 as a function of specific impulse.

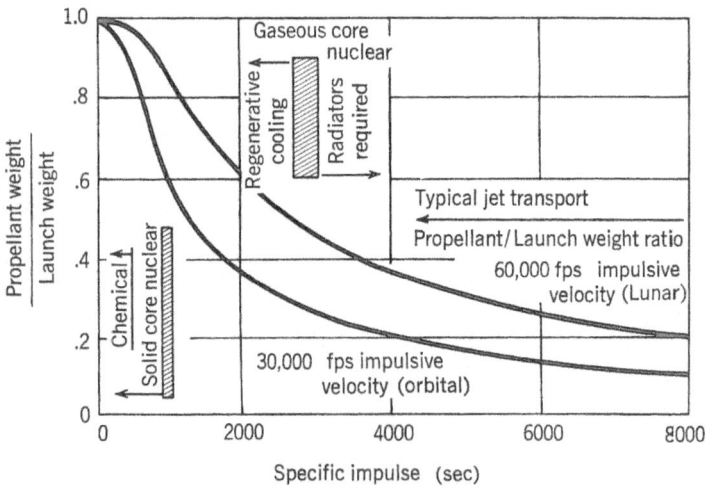

Fig. 4-23. Required fuel weights for single-stage space launch vehicles.

High-energy chemical rockets, even with specific impulse of 500 seconds would still be difficult to build as single-stage to orbit vehicles. If a solid core nuclear rocket with specific impulse of 800–900 seconds were available, orbital velocity could be achieved with 60 per cent propellant weight. A modern jet transport carries about half its take-off weight in fuel, but must carry heavy wings and landing gear. A rocket with $\lambda' = 0.85$ would be able to carry 20 per cent of launch weight as payload to orbit. It could carry 10 per cent payload to about 40,000 feet per second. Hence, nuclear rockets even with shielding can easily achieve velocity increments for orbital flight with single-stage rockets. A reusable single-stage nuclear rocket, if it could be operated with direct launch from earth, or a minimum of auxiliary boost, represents an interesting future vehicle. Obviously, should some exotic chemical with comparable specific impulse, no shielding requirements, and no earth launch danger appear, a drastic revision of orbital transportation vehicles would be in order. In Chapter 5, reusable vehicles will be discussed further.

Air-Breathing Propulsion

Normal aircraft engines operate by burning fuel with air ingested from the outside atmosphere. They do not have to carry oxidizer as a rocket does. Accordingly, their specific impulse is higher. A typical jet engine operating on kerosene has a specific impulse of about 3500 seconds. A similar engine burning hydrogen has a specific impulse of about 7000 seconds. Because of these high specific impulses, much thought has gone into using air-breathers as a means of greater efficiency for boost to orbit.

With a specific impulse of 3500 seconds, an impulsive velocity of 30,000 feet per second can be achieved with a propellant weight equal to one quarter of launch weight (see Figure 4-23). It is clear then that such a rocket would permit vehicles more like modern airplanes than current rockets in their over-all characteristics. Unfortunately, two things combine to strongly deteriorate this rosy thought.

The specific impulse achieved is a strong function of flight velocity. Air-breathing engines create thrust by increasing the linear momentum of the air, hence, the exhaust must be ejected faster than the incoming air. As the flight velocity is increased, large amounts of energy are required to create an exhaust jet with a sufficiently high percentage of flight velocity to prevent serious propulsive inefficiencies. Even with supersonic burning ram jets, which presumably can operate at any high velocity, the specific impulse decreases at high velocity to less than 1000 seconds when orbital velocity is reached. The overall effective specific impulse of an air-breathing engine, utilized all the way to orbital velocity, is not nearly as high as one would expect. It is substantially greater than a chemical rocket though, and on this basis alone, it should make an attractive propulsion system.

Air-breathing trajectories must be kept low in the atmosphere so enough air is available for the engines. In a typical rocket trajectory, the vehicle rapidly climbs vertically through the atmosphere and accelerates as rapidly to orbital velocity so that drag and gravity losses are minimized. When the air-breathing vehicle is held down in the atmosphere, it must overcome drag and support itself against gravity for a longer time. Aerodynamic heating is increased in the lower atmosphere at high velocities, and this requires increased structural weight. Wings are usually required to help the engines support the vehicle against gravity. But wings are heavy and deteriorate the mass fraction. It is difficult to find a simple way to compare pure rocket and air-breathing modes of operation. Comparisons to date indicate that the above deteriorating effects reduce the effectiveness of air-breathing vehicles to the point where they become no more effective than pure rocket vehicles for orbital transportation.

One particularly clever attempt to get around these problems is the liquid-air-cycle engine (Lace). In this system, extra hydrogen is carried and air is taken aboard and liquefied by the cold hydrogen. The hydrogen and liquid air are then used in the rocket engine to complete the operation at high velocity. This is not as helpful as it sounds. If the extra air is taken on at too low a velocity, the rocket still has most of the work to do with an unusually heavy vehicle. If the air is taken on at too high a velocity, it is too hot to be liquefied by the hydrogen. The Lace system has not yet convincingly demonstrated a performance gain which justifies its complication.

There is much more to be done in space than merely getting to earth orbit. Contrary to general opinion, getting to earth orbit is not necessarily the most difficult job to be done in space propulsion. Already we have discussed energies over 10 times that required for attaining earth orbit. In the next chapter, we shall proceed to much higher energies which would reduce earth orbital velocity almost to a minor detail.

5

Solar System Spaceships
(Velocity Approaching One Million Feet Per Second)

Celestial Mechanics

Imagine a true spaceship. Such a vehicle should be fast enough to travel throughout the solar system in a few weeks time, and economical enough so that current cost barriers would become quaint primitive estimates. To do this would require much higher velocities than have been discussed.

The previous celestial mechanics discussions contain all the essentials required for velocity regions up to one million feet per second. It is only necessary to perform the calculations. Accordingly, the celestial mechanics of this chapter will be confined to two points. One is an interesting simple way of performing calculations throughout the solar system which becomes increasingly accurate at higher velocities. The other is a discussion of certain penalties involved whenever the rocket vehicle's acceleration is small with respect to the local value of gravity. It is necessary to understand these latter penalties in order to comprehend the utilization of electrical rockets, which present a means of achieving high velocities but only at very low accelerations.

High Velocity Rockets and Gravity Fields

The previous chapter showed that as rocket velocities become higher, the perturbation exerted on their trajectories by gravity fields becomes smaller. Figure 4-2 shows that if the launch velocity exceeds only 2.3 times local escape velocity, the hyperbolic excess velocity is already 90 per cent of launch velocity. When gravitational energy is a small part of total energy, the effect of gravitational perturbation is small.

If this is carried far enough, the effect of gravity fields can be ignored. At some velocity, a spaceship will move across gravity fields so fast that negligible velocity is lost and the resulting hyperbolic trajectories will be practically straight lines at constant velocity. At the orbit of Mercury, solar escape velocity is only 222,000 feet per second. At velocities around 500,000 feet per second, then, it is reasonable to consider approximating actual heliocentric trajectories by straight lines assuming no gravitational effect of Sun or planets.

The use of straight line, constant velocity trajectories is equivalent to assuming either an infinitely high velocity or the non-existence of the Sun. The calculation of flight mechanics then would be extremely simple. We would divide the distance between the planets by the ship velocity to get the flight time, and ignore the planet's orbital motion since it would be small compared to ship's velocity (or zero if there were no Sun). This is the ultimate in high velocity assumptions, and is about as far from a Hohmann Transfer as one can get.

Since we habitually express the distance between the planets in terms of astronomical units, remembering one number will permit rapid calculation of high velocity celestial mechanics. At a velocity of 100,000 feet per second, it requires 57 days to cover one astronomical unit. Thus, the travel time is given by:

$$t_f = \frac{57 \text{ AU}}{\left(\dfrac{\Delta V}{100,000} \right)} \tag{5-1}$$

where t_f = time of flight in days; AU = distance between planets in astronomical units; and ΔV = spaceship velocity in feet per second. If all braking is applied by rocket impulse, then the right hand side of Equation 5-1 is multiplied by two.

The minimum travel time between two planets occurs when the rocket travels between the points of closest approach. For the case of rocket braking, the equation is:

$$t_t = \frac{114(\text{AU}_0 - \text{AU}_i)}{\left(\dfrac{\Delta V}{100,000}\right)} \qquad (5\text{-}2)$$

where AU_0 = distance of outer planet from Sun, and AU_i = distance of inner planet from the Sun, both measured in astronomical units.

The maximum travel time, which occurs when the rocket travels between points on the opposite sides of the Sun, is given by:

$$t_t = \frac{114(\text{AU}_0 + \text{AU}_i)}{\left(\dfrac{\Delta V}{100,000}\right)} \qquad (5\text{-}3)$$

These expressions are quite accurate at velocities of 300,000 feet per second and give a much better correlation at lower velocities than would be expected if used to calculate the total of launch and arrival velocities. Large compensating effects are present. In the actual data, velocities are higher for the interior planet involved than for the exterior. In Earth/Mercury, this difference is a factor of four. Yet these velocities average out to within 10 per cent of the simplified high velocity assumption. In the minimum travel times, the trajectories do not deviate much from straight lines. They cover longer distances than the minimum distance between the planets, but make effective use of planetary orbital velocities. Apparently these effects compensate highly.

The greatest deviations from straight lines occur for trajectories during adverse times of the synodic period when it is necessary to reach to the other side of the Sun from the launching planet. Because the actual trajectory is bent by the Sun, it travels a longer path to the other side. It is also accelerated by the Sun so that the average velocity is higher. Thus, compensating effects are present which apparently compensate far more than one would expect. Since there is little doubt that the high velocity assumption will become more accurate as the velocity is increased, it may be used with reasonable confidence as a method of calculation for spaceship design when velocities are somewhat beyond 100,000 feet per second and the total of launch and arrival velocities is desired.

With high-thrust rockets, acceleration times will not affect these calculations. A convenient number to remember is that an acceleration of one g_0 will generate 100,000 feet per second in 52 minutes. At most, a few hours will be used in acceleration or deceleration.

When Hohmann Transfers are used for interplanetary travel in the inner solar system, the velocity requirements to escape from the planets greatly exceed the heliocentric velocities required for orbital transfer. Since most interplanetary calculations have concerned the inner solar system, the traditional viewpoint is that planetary gravity fields represent the big problem in space travel. If solar system transportation systems of real convenience ever exist, the velocities required to cover the great distances of space in acceptable time are so high that this traditional viewpoint will be no longer valid. The fond science fiction dream of nullifying gravity would be of no appreciable help to a 500,000 foot-per-second spaceship. The moral is: *a rolling spaceship gathers little gravity.*

Minimum Travel Times

Minimum travel times between Earth and the minor planets and satellites of the major planets are shown in Figure 5-1. The total velocity to launch and brake at arrival is shown. Velocities of about 500,000 feet per second are sufficient to make accessible even the remotest portions of the solar system within reasonable travel times. The effect of atmospheric braking is shown in Figure 5-1, but only for Earth-Io. Comparisons of no braking and the difference between Earth to Io and Io to Earth are shown. Effects of atmospheric braking will not greatly alter the basic flight times in the high velocity region.

Synodic Period Effects

The difference in flight time between the worst and best

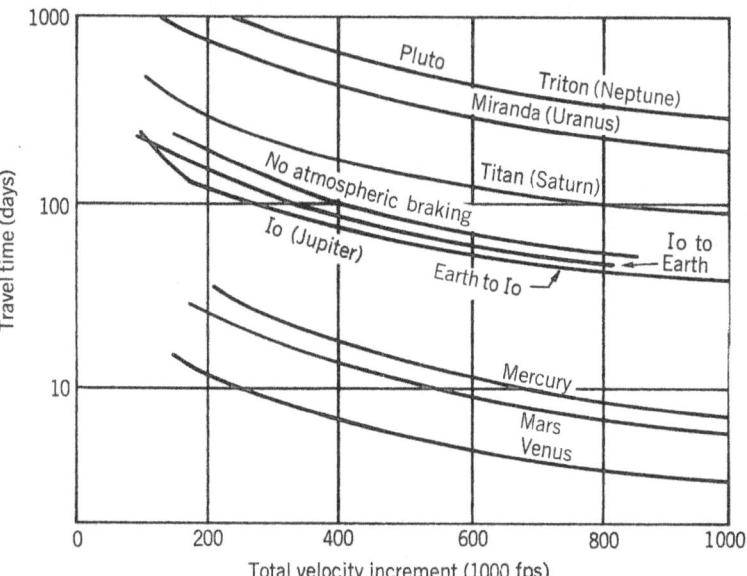

Fig. 5-1. Minimum travel times from Earth including braking re-quirements.

times of the synodic period depends only on the orbital radius of the inner planet involved, independent of the location of the outer planet (see Equations 5-2 and 5-3). This is some-what surprising at first, but not illogical when one realizes the problem during adverse times of the synodic period is that of reaching to the other side of the sun. This maneuver is likely to cost the same amount of time for a given velocity regardless of where the subsequent trajectory terminates.

For spaceship design curves, it is desirable to have a travel time which is an operational average. Consideration of the geometry of concentric circles shows maximum times will occur more frequently than minimum ones. The curves of Figure 5-2 were obtained by adding three-fourths of the time increment between maximum and minimum times to Fig-ure 5-1. Figure 5-2 represents velocity requirements which may be used for spaceship design for travel to and from earth.

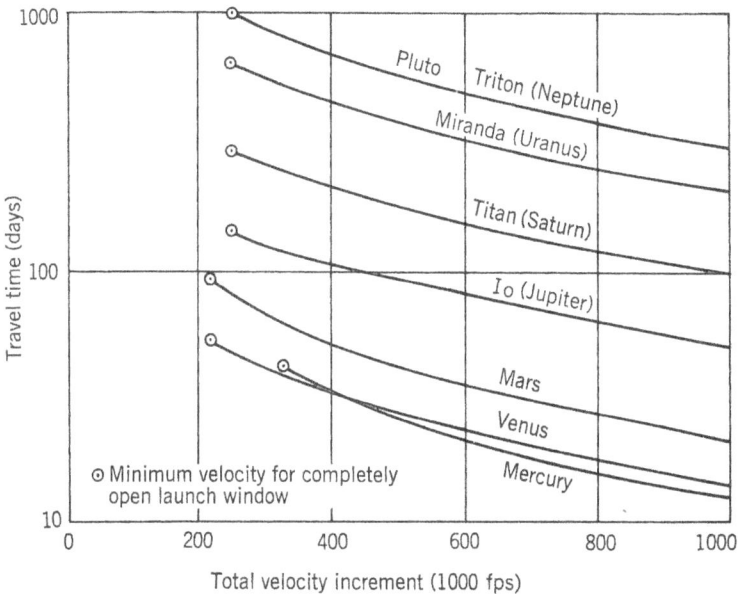

Fig. 5-2. Average travel times from Earth including braking requirements.

Planetary Bases

The curves of Figure 5-2 show that even at one million feet per second, flight time to the outer solar system is somewhat inconvenient. Furthermore, these curves give travel times between planets only if the ship can be refueled at each terminal. If it must carry fuel for the return journey, it must operate at half the total velocity shown in Figure 5-2. Except for Pluto, refueling bases at the major planets are needed much more than at the minor ones. As discussed previously in Chapter 4, bases on the larger planets' satellites would be expected to be used rather than surface bases. Figure 5-2 was drawn on that basis.

Only travel between Earth and other planets has been discussed. Travel between other planets excluding Earth may also be of interest. Use of bases in other parts of the solar system to aid in the exploration of even more remote portions should be considered. It might dictate the strategic location of bases.

At first, it would seem a good idea to use a base on a distant planet—say, Saturn—to aid the exploration of another—say, Pluto. Although intriguing, such deep bases would have restrictions. The reason is the extremely long synodic periods of the outer planets. Figure 5-3 shows synodic periods among all the planets in the solar system. The worst case, the synodic period between Neptune and Pluto, is slightly over 500 years.

The difference between travel at the optimum and at the least desirable time of the year becomes more extreme in proportion to the planet's distance from the Sun. This is illustrated in Figure 5-4 where effects of basing on selected planets are shown for a ship velocity of 500,000 feet per second. It is true that a deep space base will be closer to other deep space objects than Earth when in favorable position, but equally true that it will be much farther away when in unfavorable position. Surprisingly enough, the base wants

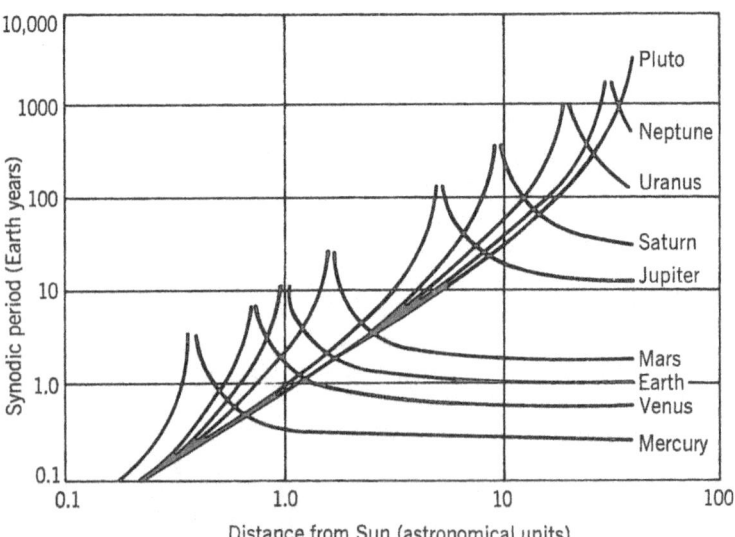

Fig. 5-3. Solar system synodic periods.

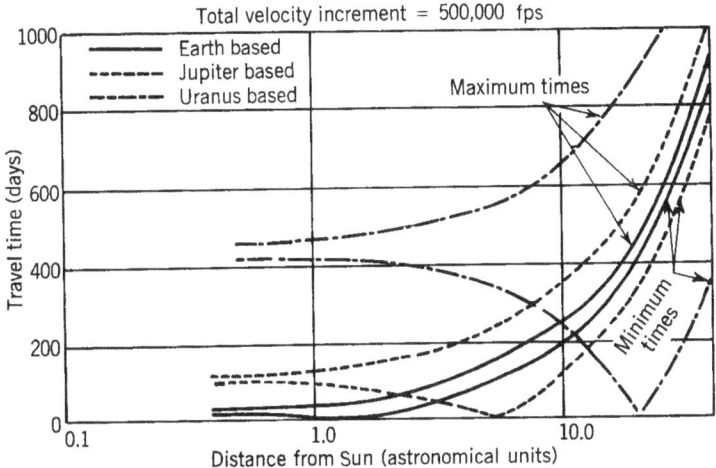

Fig. 5-4.   Travel times between planets.

to be reasonably close to the Sun, emphasizing again that the Sun is the center of the solar system. Although Mercury might be the best planetary base, Earth is sufficiently close to the Sun that it represents a good compromise. Major space logistics support operations could, from a celestial mechanics' viewpoint, be located efficiently on the Earth or its Moon. This is very convenient since the known industrial and research bases of the solar system also happen to be located in that vicinity.

The slow movements of the outer planets leads to an interesting paradox. One would assume a base on Triton to be an excellent place from which to explore Pluto, since Neptune averages 30.09 astronomical units from the Sun while Pluto will be within 33.0 astronomical units for the next 50 years. At the moment, however, Neptune is leading Pluto around the Sun. For the next 75 years, Neptune will be no more than four astronomical units closer to Pluto than Earth is. By mid twenty-first century, Earth will be closer. Due to the long synodic period of Neptune and Pluto, this will be true for more than the next 300 years.

Any ship with a heliocentric velocity greater than 138,000 feet per second at one astronomical unit will be operating at greater than solar system escape velocity. Space ships would likely operate throughout their entire lives at velocities beyond solar escape. Kinematically, they would not be in the solar system. In case of trouble, such ships would sink forever into deep space, much like an ocean vessel can be lost permanently by sinking into the ocean. If a ship could not apply braking thrust, due to engine failure, it would be lost. Conversely, one could rescue such ships or at least their crews by similar ships stationed at planetary bases.

This points out the need not only for refueling bases, but also for stationing a few ships at each base for emergency rescue operations. Such ships might be used for tankers at launch as well as rescue ships. In most discussions of inspace rescue stress is laid on trajectory compatibility and guidance training. These requirements are products of marginal propulsion systems. The ship velocity capabilities discussed in this chapter would contain adequate margin for course matching and rendezvous during rescue operations.

Escape with Low Acceleration

The desirability of avoiding the high temperatures inherent in high exhaust velocity thermal rockets has created interest in other methods of accelerating the exhaust particles. Electrical rockets are able to generate high exhaust velocities but their thrust/weight ratios are on the order of only one-one thousandth of the acceleration of gravity. Obviously, such systems cannot launch a vehicle from the earth's surface. If high in specific impulse, they may be useful in a pure space mission, even though other rockets must carry them to earth orbit. The low accelerations involved, however, require substantial modification of the celestial mechanics values presented.

The mechanism of the modification may be visualized by considering escape from a central gravity field with the vehicle initially established in a circular orbit. In high thrust rockets, enough additional velocity is added at the circular orbit altitude so that the rocket's resulting kinetic energy is equal to the potential energy required for escape from that altitude.

The escape energy required is twice the circular energy and the velocity required is $\sqrt{2}$ times the circular velocity. This calculation implies that the rocket accelerates so rapidly that all of the velocity is imparted at the original orbital altitude. Were this not so, the final velocity would be attained at a different altitude and a different escape energy would have to be achieved.

In a very low-thrust rocket, the mechanism of escape is somewhat different. Such a rocket is not able to accelerate fast enough to obtain any substantial velocity beyond orbital at any altitude. Hence, the rocket spirals slowly out through the gravity field, expending much of its energy inefficiently at altitudes higher than the original starting altitude. At each point on its outward trajectory, the low-thrust rocket must have the energy of a circular orbit at that altitude. The high-thrust rocket need not have that much energy.

For the hypothetical limiting case of infinitely low acceleration, the velocity required to escape from a circular orbit is equal to the initial orbital velocity. For the high-thrust case, the comparable number is the difference between escape velocity and circular velocity which is ($\sqrt{2} - 1$) or 41.4 per cent of orbital velocity (see Chapter 3). For a comparable initial-to-final-weight ratio, the rocket equation shows that to escape, the low-thrust rocket would require about 2.4 times the exhaust velocity of a high-thrust rocket.

Escape of an infinitely low acceleration rocket does not occur until the vehicle is an infinite distance from the gravitating source. All additional velocity inputs which contribute to the hyperbolic excess velocity are made with no benefit from the planetary gravity field. The total velocity required to escape and achieve a given hyperbolic excess velocity with negligible acceleration is given by:

$$\Delta V = V_c + V_\infty \qquad (5\text{-}4)$$

The comparable expression with high thrust, obtained by modifying Equation 4-3 for launch from initial circular orbit, is:

$$\Delta V = \sqrt{2V_c^2 + V_\infty^2} - V_c \qquad (5\text{-}5)$$

A low-thrust rocket will not have negligible acceleration (nor will a high-thrust rocket have infinitely large acceleration) and a calculation involving different manners of thrust programming and orientation is necessary to assess the actual effects. Regardless of details of the thrust program, the results are quite similar for efficient programs. Figure 5-5 shows total velocity required as a function of acceleration of the rocket for different values of hyperbolic excess velocity. This curve may be applied to all gravity fields and helps to define "low" and "high" thrust rockets. Whenever rocket acceleration is on the order of 0.001 initial gravity, almost the whole low-thrust penalty is incurred. Conversely, accelerations of about 0.5 local gravity can be tolerated without any substantial penalty unless the hyperbolic excess is large.

Solar gravity in the earth's vicinity is about 0.0006 g_o. Therefore, an electrical rocket with acceleration of 0.001 g_o

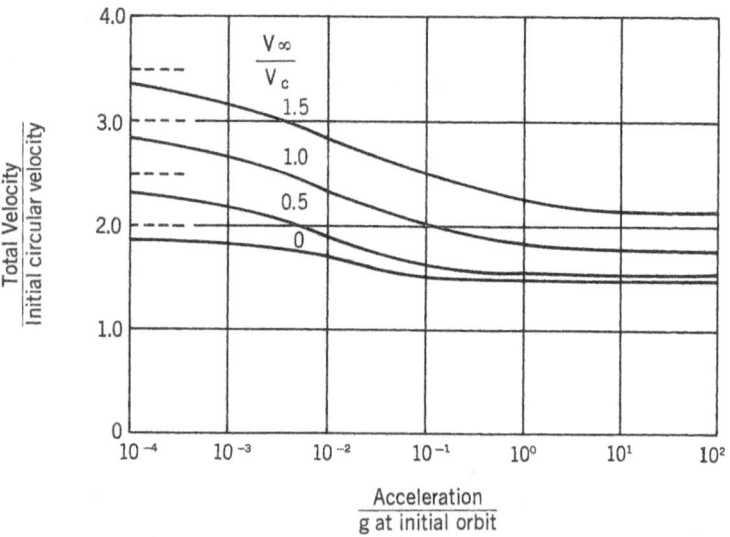

Fig. 5-5.   Total velocity to escape.

would act as a low-thrust rocket while escaping from earth, but as a high-thrust rocket in the solar gravity field. It might be effective in the solar gravity field but relatively ineffective in the earth field. It might, therefore, be boosted to higher than circular velocity by its companion high-thrust rocket. It might even be boosted beyond escape velocity. This is known as hyperbolic boost.

Flight Time with Low Acceleration

In addition to the penalties just described, low-thrust rockets suffer an added penalty if their acceleration is so low that a substantial flight time penalty is involved during the acceleration period. Assuming constant acceleration, the time required to generate velocity is given by:

$$t_b = \frac{.036}{(a/g_0)} \left(\frac{\Delta V}{100,000} \right) \qquad (5\text{-}6)$$

where t_b = acceleration time in days; and a = acceleration in feet per second^2. Note that if the acceleration is $0.0001\ g_0$, Equation 5-6 combined with Figure 5-5 gives 83 days to escape from low Earth orbit and 442 days from low Jupiter orbit. The time to accelerate penalties can be large. Equation 5-6, only valid for constant acceleration, is a good approximation if average acceleration is used.

Figure 5-6 shows the velocities required to go from Earth to various planets in the solar system as a function of travel time and acceleration of the low-thrust rocket. These curves include the velocity necessary to decelerate the low-thrust vehicle so it matches the heliocentric orbital velocity of the planet. These velocities can be used along with Figure 5-5 to synthesize requirements for low-thrust missions.

The total velocities for low-thrust missions often become extremely large. To leave Earth and establish an orbit at 1.1 planet radii around Jupiter with an average acceleration of $0.0001\ g_0$ requires 26,000 feet per second to escape from Earth, 115,000 feet per second to get to Jupiter (see Figure 5-6), and 132,000 feet per second to spiral down to low orbit around

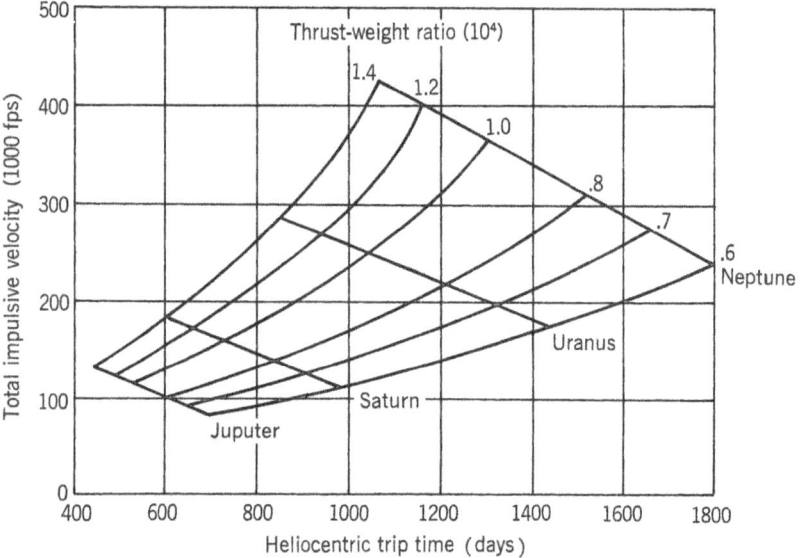

Fig. 5-6.   Heliocentric velocity requirements. (General Electric Co.)

Jupiter. The total velocity is 273,000 feet per second and the travel time is 1045 days or 2.87 years.

The comparable high-thrust Hohmann Transfer mission (orbit-to-orbit) requires only 72,000 feet per second total velocity and 2.69 years. To match a Hohmann Transfer in this case, the low-thrust rocket must produce 3.8 times the total velocity of a high-thrust rocket. Depending upon the mission, low-thrust rockets require from 2.5 to 5 times as much total velocity for a given flight time as high-thrust rockets.

Nuclear Thermal Rockets

Basic limitations on the performance of solid core nuclear rockets due to permissible temperatures in the rocket chamber were described in the last chapter. These restrictions could be removed if the temperatures could be achieved and if the nuclear fuel were in a liquid or gaseous state. If a rocket were to operate on nothing but fissioning fuel, internal temperatures would be about 100 million degrees F, and as yet, we cannot contain such temperatures. Propellant which flows through a nuclear reactor can be considered to dilute the reaction. The fission energy release is limited to the amount of thermal energy which the propellant can carry off at reactor temperature. The dilution ratio is defined as the amount of propellant used per amount of fuel burned and is basic in nuclear rocket performance.

The theoretical impulse due to nuclear energy release is shown in Figure 5-7. Undiluted effective exhaust velocity was assumed to be c/22 for fission and c/8 for fusion—both on the high side of the band given in the previous chapter, but easy to remember. The figure delineates various regions of nuclear rocket application. Both fission and fusion rockets are shown, but only fission will be discussed in this chapter. The solid core rocket region discussed in the previous chapter is shown in Figure 5-7. The middle region labelled Solar System Transportation is the subject of this chapter. The top region, Early Interstellar Travel, is the subject of Chapter 6.

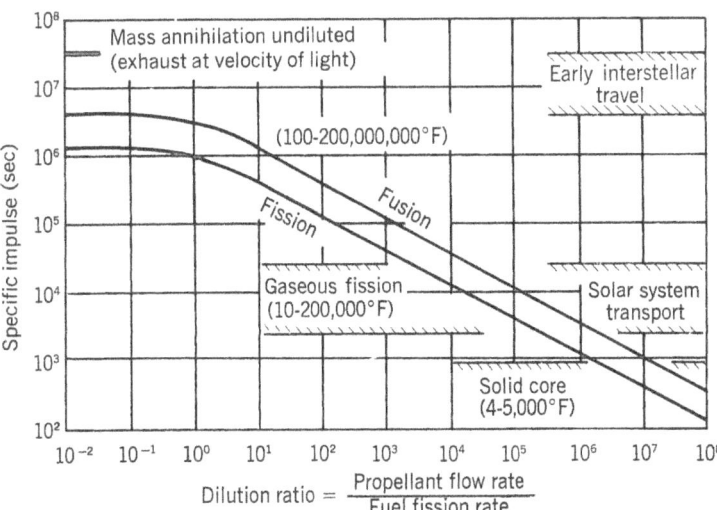

Fig. 5-7. Specific impulse from nuclear reactions.

Liquid Core Rockets

Liquid core rockets have been suggested in which a mass of liquid fission fuel is held against the outside of a rocket chamber by centrifugal force obtained by spinning the chamber. The propellant would be bubbled through this fluid and out the nozzle. In all nuclear rockets, it is desirable for safety that a minimum of fission products escape from the rocket, and for economy of operation, a minimum of unburned fission fuel escape. A high thrust requires that a large amount of propellant be put through the chamber, which tends to sweep out the fuel. Hence, liquid core reactors have a conflict between high thrust/weight ratio and low loss rates.

Although some materials have remarkably high boiling points (tungsten is about 10,500° F.), the temperature of liquid cores is limited by the boiling point of the fission fuel itself. This will restrict the maximum temperature of liquid core rockets to about 1000° F., above solid core rockets for comparable states of development. They may achieve specific impulse values of 1300 to 1500 seconds. This performance gain over current solid core rockets is substantial, but might not be worth the development effort compared to advanced solid cores. Should gaseous core rockets prove impossible to develop and should advanced carbide solid core reactors also be impossible, then liquid core rockets might become of great interest. The rest of this section will be devoted to gaseous core nuclear rockets, however, since their performance potential is so much greater.

Gaseous Core Rockets

Gaseous fission nuclear rockets have been under consideration for many years. Since the reactor consists of a gas surrounded by solid materials, it is also known as a cavity reactor. The critical mass can be made quite low in such reactors by use of a suitably large amount of reflector and moderator. Calculations show from 10 to 100 pounds depending on fuel used, moderator material, reactor and nozzle geometry, amount of poisonous material used in the pressure shell and other variables. What is practical remains to be seen. Even with low values, it is gas at a very high temperature. Its density is low, and a high pressure is required to contain the critical mass in reasonable volume. Typical internal pressures are 500 to 1500 atmospheres, or about 0.5 the pressure of guns. Gaseous core rockets tend to be quite heavy and only achieve high thrust/ weight ratios at rather high thrust levels.

Most of the earlier ideas for utilizing gaseous fission cavity reactors for propulsion involved diffusion of the propellant through the gaseous fuel so that heating occurred by direct conduction and convection. It was then necessary to separate the two gases and hopefully retain virtually all the fuel on board while exhausting all propellant. Hydrogen was assumed as propellant since low temperatures are always reassuring, even in non-temperature limited cases. Schemes such as magnetic field containment or use of centrifugal separation in some form of vortex were considered. Weight of magnetic equipment was always a problem and vortex stability and containment with any substantial diffusion rate have remained vexing. Just as with liquid cores, large thrust requires large flow through the vortex which tends to destroy the vortex.

Another family of systems has originated from these investigations. Although deceptively similar in appearance, they operate on a different principle. These systems heat the propellant by radiation from the fission plasma rather than direct intermixing. The containment problem is not one of separation but rather the prevention of mixing, a fundamentally different problem. Vortex stabilization problems certainly differ when propellant is not diffusing through the core.

A co-axial flow reactor has been suggested where a central slow-moving stream of fission fuel heats an annular, fast-moving stream of hydrogen, with separation obtained by velocity differential. This system cannot be expected to yield very good containment. Another scheme suggested is to contain the fission plasma in a quartz (or similar material) bottle. It is possible to cool the bottle to reasonable temperature while heating the propellant to high temperature by radiation, if the bottle transmits most of the radiant energy. It is like feeling the warmth of the sun's radiation through a window in winter when the window glass itself is cold to touch. Such a system would yield perfect containment and is a very exciting thought. Some of the systems suggested are shown in Figure 5-8.

The successful use of radiant heat transfer implies that enough energy can be transmitted by this process and received by the propellant. If the propellant were transparent, the radiation would go through and heat the opposite wall instead. The power output of a radiating "black" body is given by the Stefan-Boltzmann Law:

$$P_r = .501 \left(\frac{T}{1,000} \right)^4 \tag{5-7}$$

where P_r = power radiated in kilowatts per square foot; and T = absolute temperature in degrees Fahrenheit + 460. A black body is defined in physics as one which absorbs all radiation of any wavelength (color of light, infra-red, ultraviolet, radio

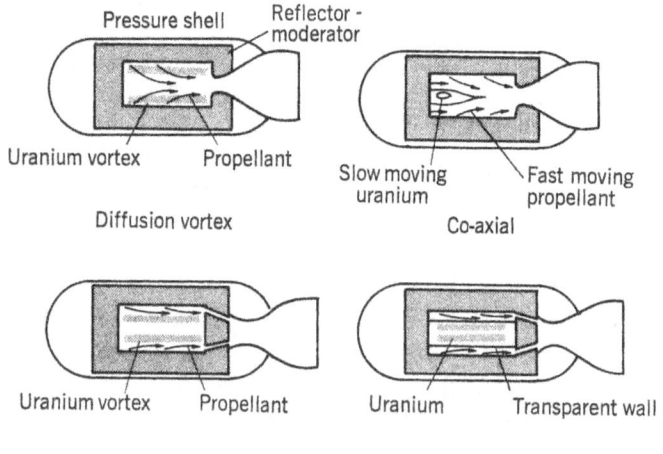

Fig. 5-8. Typical gaseous core engines.

waves, etc.). In other words, really black. When hot, a black body radiates at all wavelengths although not equally. At higher temperatures, more energy is radiated at short wavelengths. Black bodies over the complete energy spectrum are hard to find and the actual radiated power is affected by surface conditions and the back radiation from the surroundings. Equation 5-7 is a good approximation and shows the extremely important point that radiated power goes up steeply as the fourth power of the absolute temperature.

At 4,000° F., Equation 5-7 gives 200 kilowatts per square foot (268 horsepower per square foot). Although this is substantial power, it is only a small fraction of the heat transferred by convection and conduction in a solid core reactor. At 30,000° F., 431 megawatts (578,000 horsepower) are radiated from only one square foot of area. At 200,000° F., 800,000 megawatts (1.07 billion horsepower), would be radiated from one square foot of area. Once the temperature restrictions of solid core rockets are removed, it is easily conceivable that adequate heat can be transferred by radiation alone.

Depositing of the heat in the propellant requires that it must be opaque to radiation over a wide range of temperatures. Many materials including hydrogen are very opaque at moderately high temperatures and can be seeded with various materials if too transparent at low temperatures.

Fuel and Propellant Consumption

When dealing with nuclear rockets, it is useful to delineate between fuel which is burned to produce energy and propellant which is heated by the fuel energy release and constitutes most of the exhaust jet. The fuel energy release is potentially so high that, unlike chemical propellants, supplying energy is no problem. Nuclear fuels, however, only improve the rocket design to the extent that a sufficient amount of this energy can be placed in the exhaust jet to lower the rocket propellant consumption significantly.

Figure 5-7 shows propellant consumption of nuclear rockets and is valid for any material used as propellant. Using low

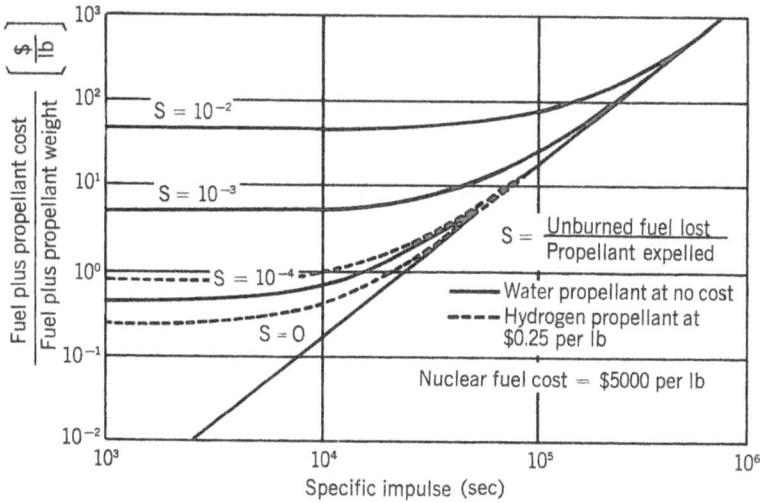

Fig. 5-9.   Cost of nuclear fission fuel and propellant.

molecular weight propellants such as hydrogen reduces operating temperatures but does not change propellant consumption. It is assumed in Figure 5-7 fuel loss rates are
sufficiently low that fuel lost has a negligible effect on exhaust
average mass flow.

Although propellant consumption is given by Figure 5-7,
fuel consumption depends directly on the degree of containment of unburned fuel. Figure 5-9 shows the effect of fuel
containment on fuel plus propellant cost for both hydrogen
and water propellants. Fission fuel releases about 17 million
times more energy per pound than smokeless gunpowder but
costs only about 5000 times as much ($5000 per pound). It is
3000 times more economical on an energy basis. So little is
consumed that at specific impulses less than 10,000 seconds,
the fuel plus propellant cost is determined almost completely
by containment. Even a loss of only one pound of fuel per
10,000 pounds of propellant expelled affects fuel plus propellant costs. With water propellant and perfect containment,
specific impulses of over 2000 seconds could be obtained for
only one cent per pound fuel plus propellant cost.

Cooling Limitations

The problem of cooling rocket engines has been referred to only briefly. It is an engineering problem rocket designers have learned to handle. Solid rocket motors simply insulate the chamber walls. Recently, liquid rockets which use techniques similar to solid chambers have appeared. They are called ablating motors and contain materials in chamber and nozzle similar to those used on re-entry bodies. They insulate the walls and also vaporize into the chamber to absorb the heat load. Such chambers have a limited burning duration before the ablator is used up.

Most liquid rocket engines use regenerative cooling. One of the incoming propellants is circulated through nozzle and chamber walls before injection into the chamber. This results in almost no decrease in specific impulse, since the energy lost through the wall is carried back into the combustion chamber by pre-heated fuel. Much skillful design is required to insure that cooling tubes are of the right size and strong enough so that heat flows into the propellant quickly enough to avoid melting the walls. Many a rocket chamber has burned through in test programs while solving these problems, but the techniques are now highly developed.

It is necessary for regenerative cooling that the propellant be capable of absorbing the heat load. Only a certain amount of propellant is available. It can only absorb so much heat before it becomes too hot to cool the chamber. This problem is increased at high specific impulse since the energy release in the chamber per pound of propellant increases as the square of the specific impulse (see Equation 1-9 and Equation 1-11).

If the propellant of a gaseous nuclear rocket is opaque enough, the thermal energy transmitted to the walls will be small. However, nuclear radiation will still heat the chamber and surrounding materials. If the incoming propellant has been raised to wall temperature, it has an enthalpy when entering the chamber which corresponds to the specific impulse of a solid core reactor at that temperature. The amount by which the gaseous heating further raises the specific impulse is given by:

$$I_{sp} = \frac{I_{sp_s}}{\sqrt{f}} \qquad (5\text{-}8)$$

where I_{sp_s} = specific impulse of propellant at temperature of solid material; and f = fraction of energy release which appears as thermally effective nuclear radiation. The value "f" is normally about 10 per cent so that a gaseous core reactor is usually limited to about 3.17 times the specific impulse of a solid core reactor if only regenerative cooling is used. The possibility of going to smaller values of "f" by means of thinner reflectors and/or relatively gamma-transparent shells is interesting. Only the neutrons must be reflected, and hence their energy absorbed in the reflector, in order to contribute to reactor criticality. It is logical to balance the neutron reflective properties of materials with their relative gamma transparencies and thermal cooling properties to give optimum reflectors for these applications.

The specific impulse of a gaseous core rocket may be extended beyond the limit set by regenerative cooling if a space radiator is used to reject excess heat. The major problem then becomes the weight of the radiator. As of 1965, most investigations of radiator configurations for nuclear propulsion have centered around the requirements for nuclear electric systems for either propulsion or auxiliary power. These requirements are totally different from those for gaseous fission rockets. Nuclear electric systems must be designed for long operating times (years) so that such problems as meteoroid penetration of the radiator surface must be considered in terms of long-time probabilities. This strongly influences radiator weight. A high-thrust gaseous fission system would only operate for periods of hours at a time and, therefore, the use of short-life radiators is pertinent. It is true that the radiator must survive for the total flight duration, not simply the engine burning period, since the engine must be used for braking at arrival. However, total flight times will be much shorter than for electrical systems, the radiators might be protected while not radiating, and the loss of a radiator segment would not be very crippling.

Considerably more important, the radiator temperature of a gaseous fission system can be as high as it is possible to build radiators, since it need only prevent the engine from melting. In a nuclear electric system, a balance must be struck between the efficiency of the conversion process, which requires the rejection of heat at a low temperature, and the decrease of radiator weight which, in general, occurs at high temperature. Radiators usually want to operate at about 0.75 of the maximum cycle temperature. The maximum cycle temperature is determined by the ability of either rotating machinery or thermionic systems to operate for periods of years. Radiators for gaseous fission rockets should be operated at much higher temperatures than those for nuclear electric systems, and might be easier to design because of the vastly shorter operating time.

If one considered a gaseous fission nuclear rocket of one million pounds thrust operating at 2500 seconds specific impulse and 75 per cent efficiency, Equation 1-11 gives 97.5 million horsepower or 72,900 megawatts as the internal energy release. If 10 per cent of this were rejected by a radiator operating effectively at 4000° F., Equation 5-7 shows about 36,000 square feet of radiator area required. At one pound per square foot, the radiator would weigh only 3.6 per cent of the thrust. If a 1400° F. radiator temperature were used, 34 times as much radiator area would be required, and the radiator weight at one pound per square foot would equal the thrust. High temperature radiators will likely be heavier than low ones, but not by a factor of 34.

If the 72,900 megawatt reactor of the previous paragraph were operated at high specific impulse with radiators, its thrust would be less than one million pounds. It would be inversely proportional to specific impulse (see Equation 1-11). The actual weight of such an engine is difficult to estimate. Figure 5-10 shows the variation of thrust/weight ratio with I_{sp} for a gaseous nuclear rocket system with radiator using as a basis an assumed thrust/weight ratio of 20 at an I_{sp} of 2500 seconds. It might require more than one million pounds thrust to achieve this value. Both hydrogen and water are shown as propellants. The thrust/weight ratio falls off substantially at high specific impulses but is greater than one to beyond 10,000 seconds.

Use of water, ammonia, or other non-hydrogen working fluids should be seriously considered in gaseous fission engines. Not only are better ship designs permissible due to small tankage sizes and ease of propellant storability, but use of a higher density propellant might ease the fuel containment problem if a vortex system were used. If so, it could result in smaller, lighter engines. It might also result in earlier development programs if the ability to prove adequate containment occurred at an earlier time.

A result shown in Figure 5-10 is that the value of thrust/weight ratio is independent of propellant as the specific

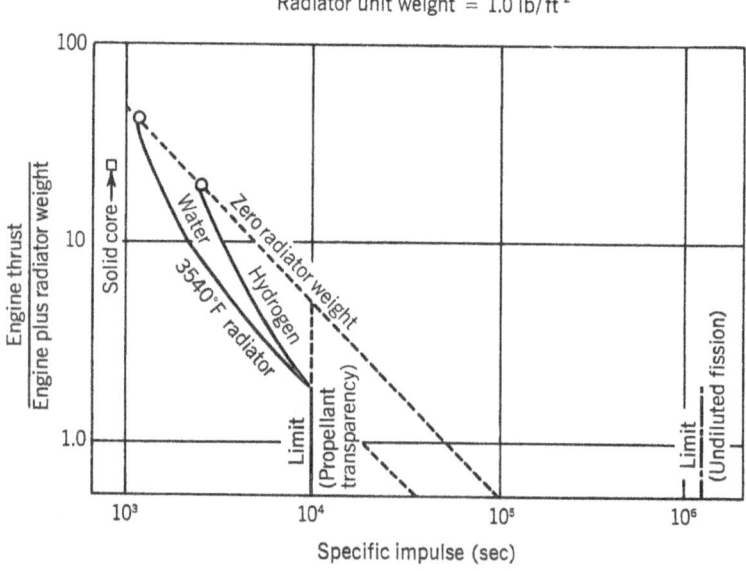

Fig. 5-10. Thrust/weight ratio of gaseous fission engines.

impulse approaches 10,000 seconds. This is because the reduced propellant flow at high specific impulse results in such
small thermal capacity in the incoming fuel that the engine
must be almost completely cooled by the radiator system. The
engine uses the same amount of energy to generate a given
specific impulse, the same fraction of energy must be rejected by
the radiator, and the radiator area is hence unaffected by the
type of propellant used.

An engine design cooled by radiator alone should be investigated. Such an engine might be easier to develop since a
major interaction between propellant and cooling system
would be severed. Furthermore, it might more easily use a
variety of propellants. This could be very helpful in early
planetary exploration.

A limitation on specific impulse of 10,000 seconds has been
shown in Figure 5-10. This is due to an unfortunate tendency
of propellants examined. Although adequately opaque to absorb the radiant energy at medium-high temperatures, they
apparently become transparent at very high temperatures.
Seeding the flow, which is effective at low temperatures, is not
promising at high temperatures. Other limitations may occur
at lower specific impulse. When the flow expands through a
nozzle, the propellant may become transparent as it cools and
make it impossible to cool the nozzle beyond a certain expansion ratio. These limitations are not yet well enough understood to know which, if any, are fundamental.

A number of unknowns exist with respect to gaseous fission
engines, some of which are mentioned in the previous para-
graphs. Some, like nozzle problems, are due to these engines
being the most powerful ones yet considered in this book. As
long as propellant was supplied, a 3 million-pound thrust
gaseous fission engine of 2500 seconds specific impulse would
release 300 million horsepower, six times that of the 16-inch
naval gun. At 220,000 megawatts, it would be 200 times as
powerful as current solid core rockets. *If the problems of
such engines are impressive, it is because we are finally
discussing the class of energy control which could make true
spaceships possible.*

Nuclear Electric Rockets

Electrical rockets should be attractive. If the exhaust is accelerated by electrical means, high velocities can be generated without resorting to high temperatures. Because of this, a great deal of thought and effort has been given to electrical rocket development.

Three principal classes of electrical rockets have been considered with a number of differing detailed implementations in each group. Electrothermal engines consist of an electrical means of heating a propellant. In the resisto-jet, hot tungsten plates heat a propellant passed over them. In the arc-jet, an electrical discharge is passed through the propellant heating it to higher temperatures than can be attained by normal heat transfer from solid surfaces. Utilizing hydrogen as propellant, resisto-jets have produced over 1000 seconds specific impulse, and arc-jets have produced 1500 to 2000 seconds. Electrothermal engines are not a means of circumventing the temperature problem, but rather a way of producing higher temperatures.

Higher specific impulses may be attained by electrodynamic or magnetohydrodynamic (MHD) engines. In this case, a gas is ionized at moderately high temperatures, then accelerated by magnetic fields. A number of thrust units of this sort have been built. Some have been operated at specific impulses as high as 10,000 seconds.

Electrostatic or ion engines are also suitable for high specific impulse. These engines use electrostatic fields to accelerate charged particles with no assist from fluid flow forces such as in MHD engines. Ion-thrust units also have been built and tested up to 10,000 seconds specific impulse and should be extendable to higher specific impulse. The first electrical rocket flight test, an ion engine of .006 pounds thrust, occurred at Wallops Island, Virginia on July 20, 1964. Figure 5-11 shows typical engines in all three categories.

So far, electrical rockets seem very good, as if they are the answer to everything. There is, however, a "catch" and its name is electrical power supply. Electrical rockets take a source of energy—chemical, nuclear or solar—and convert it

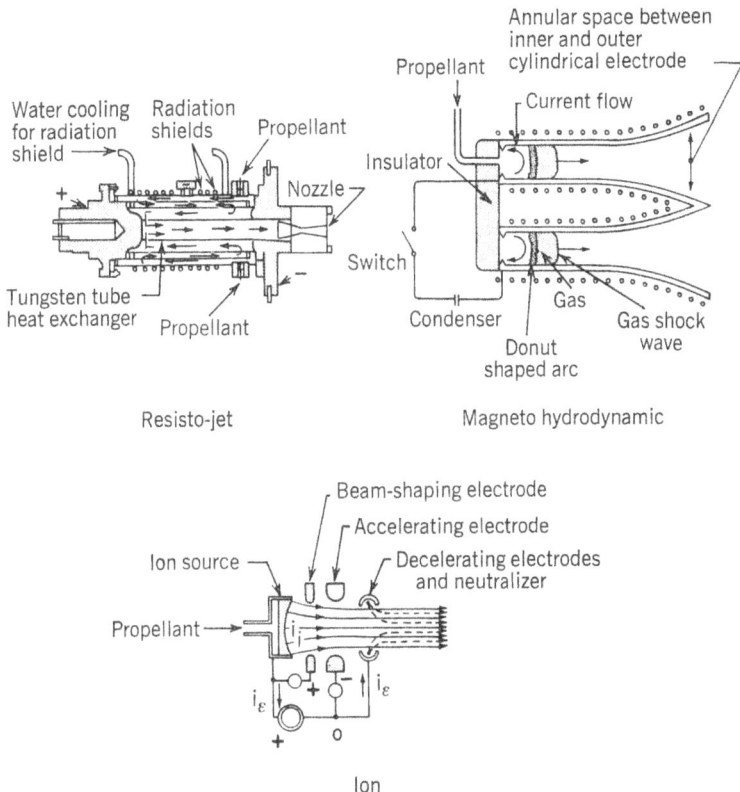

Fig. 5-11.   Types of electrical rocket thrusters.

into electrical energy for use in the accelerating apparatus. The electrical rocket, like ship, airplane, or automobile, must handle all the energy of the propulsion process. Hence, it loses the main energy manipulating advantage which thermal rockets share with guns. The power to produce one pound of thrust with 75 per cent energy efficiency is 29.1 kilowatts (39 horsepower) at 1000 seconds I_{sp} and 291 kilowatts (390 horsepower) at 10,000 seconds i_{sp} (see Equation 1-11). Terrestrial internal combustion engines run about one pound per horsepower without any electrical power generating equipment attached. It is clear vast improvements in power conversion weights are required for high thrust/weight ratio.

The large energy requirements associated with high specific impulse mean that chemical power supplies are useless for electrical rockets. Only solar or nuclear power is of interest, except for thrustor test missions. Much development is under way on nuclear electric power supplies for electrical power needs and as power units for electrical rockets. A typical 1965 technology system—SNAP 8—weighs 175 pounds per electrical kilowatt. Design targets for advanced systems are 10–20 pounds per electrical kilowatt. Studies of very advanced systems have sometimes led to weight estimates as low as two pounds per electrical kilowatt but rarely lower. The weight of power supply dominates nuclear electric rocket performance. If gaseous fission engines do achieve a thrust/weight ratio of 20 at an I_{sp} of 2500 seconds, they will be effectively converting power to thrust at the rate of about $\frac{1}{1500}$ pound per exhaust kilowatt which is over 20,000 times better than current electrical rocket design objectives without even allowing for thrustor inefficiencies.

Electrical Rocket Performance

If α is power plant specific weight (pounds per exhaust kilowatt), then total energy output of the power plant per unit weight is t_b/α. If this could be converted into kinetic energy of the power plant (see Equation 1-4), the resulting velocity would be:

$$V_{ch} = 64,100 \sqrt{\frac{t_b}{\alpha}} \qquad (5\text{-}9)$$

where V_{ch} = characteristic velocity in feet per second; t_b = power plant operating time in days; and α = power plant specific weight in pounds per exhaust kilowatt. The characteristic velocity is a great help in understanding electrical rockets. It represents the limiting velocity achievable with 100 per cent efficiency and zero payload. Since it increases with burning time, electrical rockets can be made efficient by extending the burning time. Unfortunately, only the rocket efficiency improves with long burning time. Everything else gets worse.

Chemical rockets have a fixed maximum exhaust velocity for each propellant combination. One cannot concentrate the energy release into less propellant. Electrical rockets have an energy source which is separate from the propellant, and can operate at any exhaust velocity for which efficient thrustors are available. Since the total energy output of the power supply must appear as exhaust energy, we have for perfect efficiency:

$$w_\alpha V_{ch}^2 = w_p v_{ef}^2 \qquad (5\text{-}10)$$

where w_α = weight of power supply in pounds, and w_p = weight of propellant in pounds. Thus, the exhaust velocity can only be increased with respect to characteristic velocity by devoting less of the total power plant weight to propellant. Clearly there will be an optimum value for exhaust velocity.

The rocket equation can be derived in terms of characteristic velocity with optimum exhaust velocity program. The result is shown in Figure 5-12 which bears a strong resemblance to Figure 1-6. Figure 5-12 can be used with Figures 5-5 and 5-6 to estimate electrical rocket performance if the power plant specific weight is known. The power plant operating time may be taken as two-thirds of the total travel time, since trajectory analysis tends to show one-third acceleration, one-third coast, and one-third deceleration as roughly optimum burning programs. The optimum exhaust velocity program requires that exhaust velocity always equal flight velocity.

The nature of the problems of nuclear electric rockets may be illustrated by an example. If we wish to go from low Earth orbit to a Saturn orbit at the radius of Titan, then Figures 5-5 and 5-6, Equation 5-6, and the planetary data of the table on page 82 show a velocity of 210,000 feet per second required at 950 days travel time and 290,000 feet per second required at 600 days. The characteristic velocity with $\alpha = 20$ pounds per kilowatt is 362,000 feet per second in the former case, but only 286,000 feet per second in the latter. Hence, at shorter travel

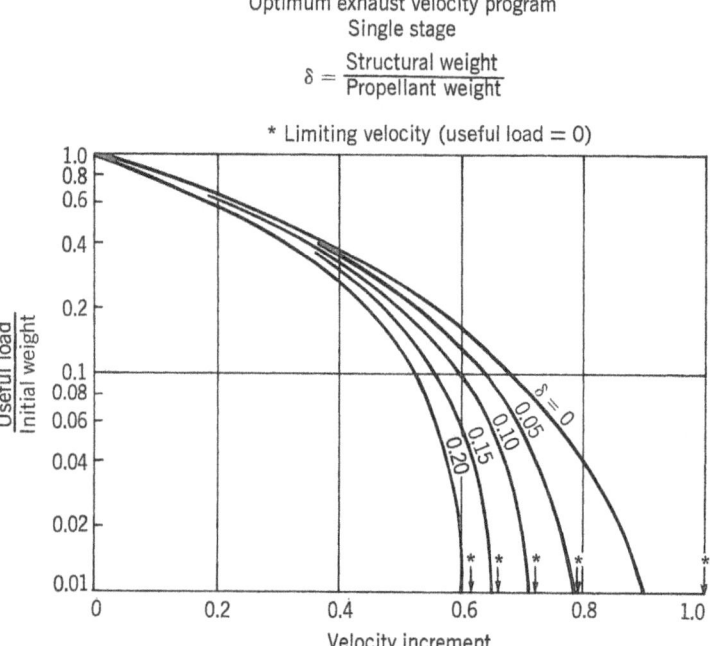

Fig. 5-12.   Electrical rocket performance.

times, the need for increased power cannot be met unless α is decreased.

Electrical rockets give large payloads at long travel times but fall off rapidly at short times since the velocity requirement goes up at the same time the characteristic velocity goes down. If specific power is around 40 pounds per kilowatt, they are only marginally competitive with solid core nuclear thermal rockets. At 10 pounds per kilowatt, they still do not compete well with gaseous core nuclear rocket possibilities. Any comparison of the relative worth of nuclear electric and nuclear thermal rockets usually deteriorates to a discussion of relative operational dates for an electrical power supply of given specific power as compared to a nuclear thermal rocket of given specific impulse.

Solar Powered Rockets

Use of solar energy to power rockets has been considered often and has been utilized for attitude control. A solar electric power supply can be used instead of a nuclear supply to power any type of electrical rocket discussed. Even though it can give a human a severe sunburn, solar energy at the distance of the Earth's orbit around the Sun is weak. Solar energy falling on one square yard is about 1.12 kilowatts. To collect the amount of energy (1100 megawatts) generated by the Nerva nuclear rocket, one million square yards of surface would be required. This is two-thirds of a mile in diameter. Because solar flux is so low, any solar powered rocket system will be of low thrust/weight ratio.

One interesting solar-powered rocket idea is the solar sail. Radiation pressure from the Sun is caught in a sail and used for propulsion. This is exactly analogous to ocean sailing vessels, including the ability to tack against the "wind." It is attractive for the same reasons. All the power used is free. It has the same disadvantages. It incurs the longest travel times. The radiation pressure field of the Sun at the distance of the Earth requires over one million square feet of perfect reflector to generate one pound of thrust. The "solar wind" of high velocity protons exerts only 0.001 of the pressure of the Sun's radiation and, in any event, would penetrate thin material with ease. It will not be easy to find convincing uses for "Sunjammers" if the promise of gaseous fission powered spaceships materializes.

All solar-powered propulsion systems behave differently from self-powered rockets. They depend on the Sun's energy. This energy varies inversely as the square of the distance from the Sun. Solar-powered devices, therefore, will intercept more energy and work better closer to the Sun. Mercury is about one-third the distance from the Sun that the Earth is, and a solar-powered device would be 10 times as powerful at Mercury. The acceleration of a solar rocket, however, would vary with distance from the Sun much in the same way the Sun's gravity field varies. If it were a "low-thrust" rocket at Earth, it would be the same throughout the solar system.

We are interested also in going outward from the Earth. At Jupiter, which is roughly five times Earth's distance from the Sun, the energy of a solar device would be only $\frac{1}{25}$ th as great as at Earth. At Pluto, it would be weaker than at Earth by a factor of 1000. Solar-powered rockets, which at best are marginal for use in the vicinity of Earth, would be expected to be totally inadequate for use throughout the entire solar system.

Nuclear Pulse Rockets

Nuclear pulse rockets make use of small nuclear bombs as propulsive devices. They give promise of high specific impulse and relatively high thrust/weight ratio. The internal pulse engine detonates a series of bombs in succession in a large rocket chamber. For each detonation, a suitable amount of propellant is placed in the chamber around the bomb and is heated and pressurized by the bomb energy release so that it expands out of the nozzle. Calculations of such systems have been performed, but no serious development has been started.

External pulse rockets detonate bombs behind the ship. They are arranged to throw propellant against a suitably shock-mounted striker plate. Cooling restrictions are different from gaseous core nuclear rockets since the plate receives only a sharp thermal pulse which can be radiated away between detonations and much of the nuclear radiation can be absorbed in the propellant. The Orion program has performed a number of chemical high explosive simulations of a nuclear pulse system, but no nuclear tests have been performed as of 1965. The status of nuclear pulse rockets with respect to any nuclear test ban is a continuing subject of discussion.

Although nuclear pulse rockets are interesting and may be used in the future, we will not discuss them in detail here. The fuel cost of any bomb propellant system would seem to be much higher than that of a well-contained gaseous fission system. Each detonation, of which thousands are needed, requires the assembly of a critical mass, all of which is lost. It is doubtful if pulse systems could ever compare economically with gaseous fission unless almost pure fusion bombs were used. Should the latter become possible, nuclear pulse rockets might become highly competitive.

Spaceship Design Philosophy

The performance capability of gaseous fission engines is so high that their manner of utilization may be completely different from other rocket engines. They should permit the practical realization of a true spaceship for the first time. The performance capability is so high, it is reasonable to consider a ship design which can operate throughout the entire solar system.

Some possible design interactions between the ship and its propulsion system, as well as some other operational considerations, will be presented here. Other propulsion systems, such as a nuclear pulse system based on clean fusion bombs, or a nuclear electric system with specific power much less than one pound per kilowatt, might also achieve some of the advantages discussed. We will consider only the gaseous fission system, however, since it seems to combine most advantages in one package.

Versatile Ship Design

Under certain circumstances, much can be learned about spaceship design without detailed knowledge of the missions to be performed. When considering total solar system transportation, we face a variety of missions. Which will be paramount is not clear. One approach to the problem is to present the characteristics of the vehicle as a function of total velocity increment which the ship can achieve. This is exactly analogous to the aircraft design practice of presenting operating characteristics as a function of range. In this way, the ship's ability to deliver payload to a certain velocity economically can be easily understood. Complex mission analyses can then be made to reflect the maximum design velocity increment required.

Fuel and Propellant Costs

In understanding spaceship operating costs, it is instructive to consider first only fuel and propellant cost. This cost represents the minimum achievable. It is particularly important to understand how to achieve a low fuel and propellant cost when truly reusable ships are employed. In transport aircraft practice, reuse is so high that initial airframe costs are only a small fraction of the operating cost, and fuel costs represent about one-half of the total. We shall examine fuel costs for their basic limitations on performance, then see how closely these limits can be approached with reusable ships.

Fuel and propellant costs as a function of total velocity increment for chemical, solid core nuclear, and gaseous nuclear rockets are shown in Figure 5-13. A number of different propellants and degrees of containment are shown for gaseous fission engines using the fuel costs of Figure 5-9. All curves are for single-stage ships. A λ' of 0.85 was assumed. This corresponds to an advanced reusable vehicle design. In addition, the payload was taken as 0.85 useful load to make an allowance for the structure containing the payload.

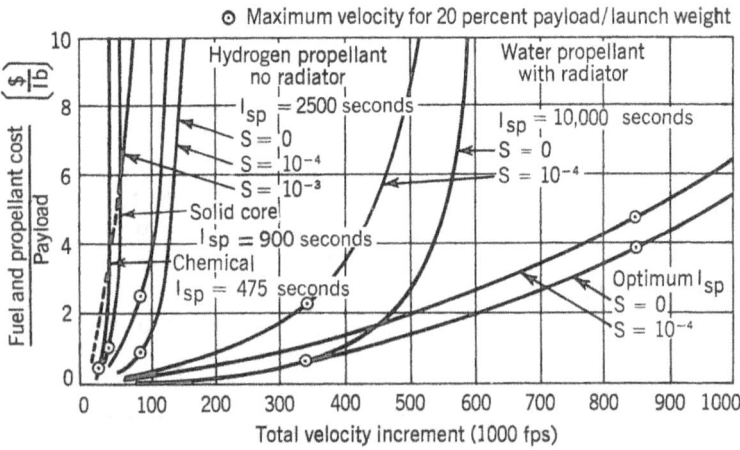

Fig. 5-13. Single-stage spaceship fuel and propellant costs.

On a fuel and propellant cost basis alone, a gaseous fission engine without radiators and with separation ratio of 10^{-3} is not significantly better than a solid core engine. Gaseous engines with better containment would be much better. Gaseous engines with space radiators, but with specific impulse limited to 10,000 seconds, can drive ships up to about 500,000 feet per second and still maintain reasonable fuel cost. The attainment of a fuel separation ratio of 10^{-4} is almost as effective as perfect fuel containment.

The optimum fuel cost curves for gaseous fission engines with radiators were obtained by determining the optimum specific impulse for each velocity and separation ratio. This is necessary since too low a specific impulse will result in excessive propellant cost, while too high a specific impulse will result in excessive fuel cost. The optimum specific impulse is much higher than 10,000 seconds for all velocities beyond a few hundred thousand feet per second. Hence, these curves represent a future capability presently unattainable due to propellant transparency problems at high temperatures. Were this not so, gaseous fission ships could be driven to almost one million feet per second before fuel costs became a limitation.

Structural Cost Amortization

Future orbital transportation costs are usually estimated at several hundred dollars per pound. With hydrogen-oxygen as propellants, an advanced orbital rocket would use less than 13 pounds propellant per pound payload. Since hydrogen-oxygen costs less than 10 cents per pound, the cost of propellant is about $1.00 per pound payload. Expensive equipment and large launch costs result, however, in total costs 100 to 1000 times the cost of the energy used.

The obvious key to reducing these numbers is reuse of the equipment, as with transport airplanes or any other transportation device. It is the ammunition business, not transportation, which thrives on throw-aways. Many analyses of reusable rockets result in only modest improvements from current practice. The reason is shown in Figure 5-14. Vastly

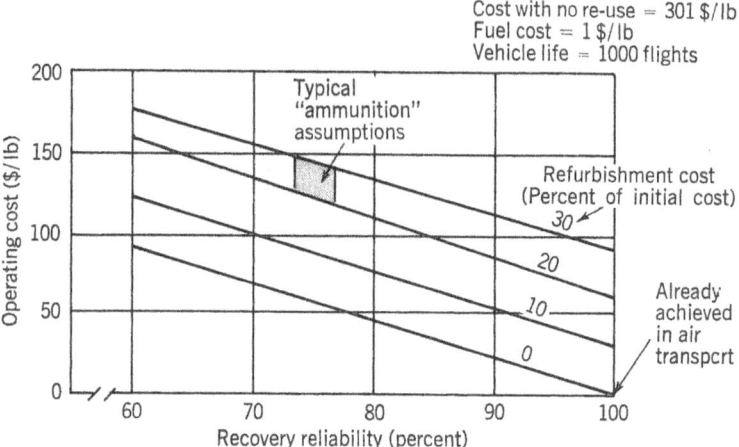

Fig. 5-14. Transportation versus ammunition re-use assumptions.

different assumptions as to recovery reliability and refurbishment (maintenance) cost characterize ammunition and transportation assumptions. Typical ammunition assumptions are recovery reliabilities of 75 per cent and refurbishment costs of 25 per cent. After spending that much on refurbishment with that low a recovery reliability, an improvement of two in cost is the best to be expected.

Also shown in Figure 5-14 are values already achieved in air transportation many decades ago. Recovery reliability is so close to 1.00 that it cannot be seen on this scale. The same is true of maintenance cost, which is about 0.04 per cent. A transport airplane is more complicated than a ballistic missile. After one lands, however, a few people turn it around, give it some fuel, pat it on the head, and it takes off again. This should be the goal for future spaceships. We shall estimate the results of such transportation operations.

A transport aircraft is used about 50 per cent of the time. Average flight durations are less than four hours. Such vehicles are hence used over 1000 times per year. Space travel durations are much longer, though, and interaction between travel duration and number of reuses must be considered.

For lunar missions, large numbers of reuses can be envisioned. Typically, 100 flights per year (50 each way) would occur on the basis of two-day travel times, one day turn around at each terminal, and allowing Sundays plus two weeks off for vacation. Over a ten-year ship lifetime, 1000 uses will be achieved.

To obtain the number of interplanetary uses, one must assume a ship total life. Typically, transport aircraft are designed for 40,000 hours (4.6 years) total life. With slightly less than 50 per cent utilization, such a vehicle would last for 10 years. They always last much longer but the amortization time of the airframe is usually about 40,000 hours, since new equipment always becomes available in even shorter time.

Selecting a suitable lifetime for a spaceship presents a technical dilemma. One viewpoint would take ten years as above. An even shorter lifetime might technically be justified due to severe loads associated with atmospheric entries and the generally unknown operational environment of space.

It may be, however, that spaceships will last longer than transport aircraft. The transport main propulsion system operates continually during flight, and it continually faces the temperatures and gust loads within our atmosphere. The question is whether spaceship operating life should be determined by the total time of operation, or only by the times during which the main engines operate and/or it is within an atmosphere. In other words, is a spaceship coasting between planets actually operating in the transport aircraft sense, or is it merely parked in space, breathing quietly, waiting for its next mission?

One can make a case for the latter point of view in terms of the space environment that the ship faces while coasting. The ship would have to be on interplanetary runs for several centuries in order to build up 40,000 hours of engine and atmospheric operation. It is bound to be replaced, however, by better equipment within a few decades. As a base for calculations, we shall assume 25 years as useful ship lifetime.

Various spaceship weights versus velocity are shown in Figure 5-15 for specific impulse limited to 10,000 seconds and for optimum specific impulse. These curves are for versatile ships

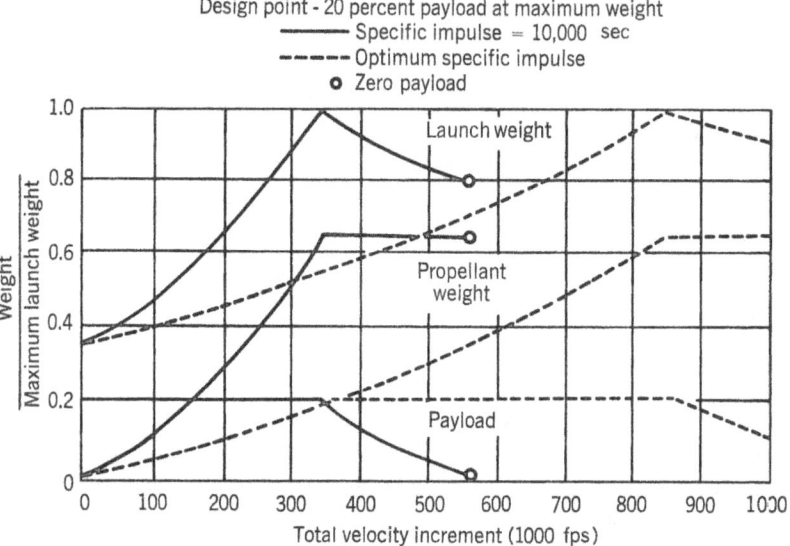

Fig. 5-15.   Spaceship payload capability.

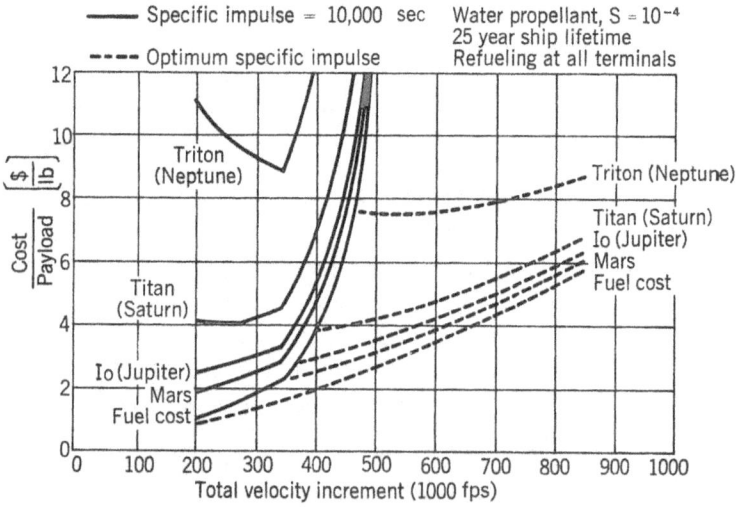

Fig. 5-16.   Single-stage spaceship fuel, propellant, and structure costs.

designed for 20 per cent payload, then operated by off-loading propellant at less than design velocity and off-loading payload at greater than design velocity, just as transport aircraft.

By using the travel time data of Figure 5-2, the weight data of Figure 5-15, the fuel plus propellant cost data of Figure 5-9, and assuming a vehicle cost of $100 per pound, the curves of Figure 5-16 were obtained. The currently estimated cost of a supersonic transport is $100 per pound. These curves show fuel plus amortization of spaceship cost as a function of velocity increment for the missions selected.

The lowest curves on Figure 5-16 are fuel cost only. For operations as far as Saturn, structural costs are comparable to fuel costs. Further improvements in convenience of operation could be achieved with engines not limited to 10,000 seconds specific impulse. In that case, velocity increments beyond 500,000 feet per second would be economically reasonable.

The average travel time between planets corresponding to the velocities of Figure 5-16 is shown in Figure 5-17. With specific impulse limited to 10,000 seconds, the travel time to

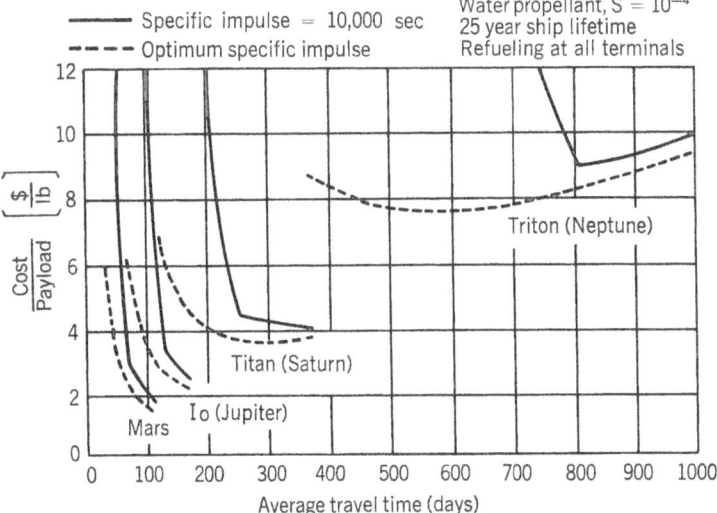

Fig. 5-17. Single-stage spaceship fuel, propellant, and structure costs.

Jupiter is under four months. Inner solar system travel times need not exceed two months. Optimum specific impulse is more important at Saturn and beyond.

It must be realized Figures 5-16 and 5-17 represent certain assumptions. Better or worse situations may occur. A specific impulse of 10,000 seconds requires gaseous fission engines with radiators, and most 1965 thinking concerns engines without radiators. In that case, the velocity increment would be only 25 per cent of the values shown. The structural weight may be heavier than assumed. Furthermore, the economic penalty of possibly ejecting a critical mass of fuel in the process of shutting down the engine has not been included. This will be about $100,000 per shutdown.

On the other hand, perfect containment might be achieved. We might design ships for each velocity increment, rather than use the single design assumed. One can get a greater utilization of vehicles by refueling those which go on deep space missions. This may be preferable to multi-stage vehicles, since a fleet of ships used for refueling can also be used for other missions. No attempt will be made here to present detailed effects of refueling. Cursory checks show that over 200,000 feet per second can be added for reasonable cost with two refuelings.

The greatest conservatism of all in Figures 5-16 and 5-17 is in the magnitude of the ordinate scale. Costs beyond $12 per pound have not been shown. The entire set of curves is about 100 to 1000 times lower than virtually all space cost analyses to date. *This is the lure of the gaseous fission powered spaceship-transportation costs throughout the solar system which are not much greater than terrestrial transportation costs.* It cannot be done with chemical energy. Assuming the best current chemical rocket, hydrogen-fluorine with exhaust velocity of 15,000 feet per second, the entire weight of the Earth would be required to accelerate one pound to 750,000 feet per second. The weight ratio of 10,000 occurs at only 120,000 feet per second. At one million feet per second, kinetic energy per pound is about 740 times that at Earth escape velocity, and about 9000 times the energy release of smokeless gunpowder.

It is not surprising that only nuclear rockets can go economically to these velocities. No matter how often refueled, chemical rockets cannot compete with real spaceships.

Transportation Development Philosophy

Transport aircraft not only achieve high reuse, they have sufficient redundancy to permit flying with partial equipment failures, and to abort successfully from any flight condition. This last capability is very important to any development of spaceships. It is sometimes thought that recoverable vehicles would be more expensive to develop since they are more complicated. This might be true of the recovery of marginal performing rockets, but not of a properly designed reusable spaceship.

One cannot over-emphasize the effect of this difference in development philosophy. Commercial transports are extensively tested and much equipment is refined by flight tests. They become reliable vehicles for expenditures small in space budget terms because it is possible to test over and over for reasonable expenditures. If high performance propulsion systems can, for the first time, permit us to pursue a space vehicle program with the development techniques of transport aircraft systems, the resulting impact on development programs should be tremendous.

This difference in development philosophy has existed from the very beginning. The Wright brothers made four flights on the first day, but it was eighteen days before Goddard fired his second liquid rocket.

The ability of a reusable ship to be tested conveniently depends on how close to the ground the main engine can be ignited, and the state of the ship after engine operation. In these items, gaseous fission engines may be better than solid core nuclear engines. Gaseous fission powered ships should be able to dispose of their fission products safely in deep space but solid core engines continually build up fission products internally as they are used. Hence, a gaseous core ship may return to earth with no fission products aboard if it uses aerodynamic braking, a degree of cleanliness which cannot be achieved with solid core engines. If it is constructed of properly selected materials, the radioactivity induced in the structure by neutron bombardment from the engine will be almost negligible.

Figure 5-18 shows the ground dose experienced in case of fission product ejection versus ignition for gaseous nuclear engines. Since a single acute dose of 25 rems produces no ill effects in humans, Figure 5-18 shows that only 5000 feet is sufficient to meet very stringent safety numbers even in a catastrophe. If the first 5000 feet were obtained with chemical boost, about seven per cent additional launch weight would be necessary. This is quite reasonable. A reusable gaseous fission ship might be flight-tested safely without most of the elaborate test procedures of chemical rockets. The fact that its performance is far higher has nothing to do with its basic safety.

During each launch of a 500-ton spaceship from earth, the fission products generated at altitudes of less than 200,000 feet are about equivalent to those of a three kiloton bomb. Fission products equivalent to a 7.5 kiloton bomb will be generated by the time orbital velocity is reached. (This number is easily checked by using 10.43 million foot-pounds per pound

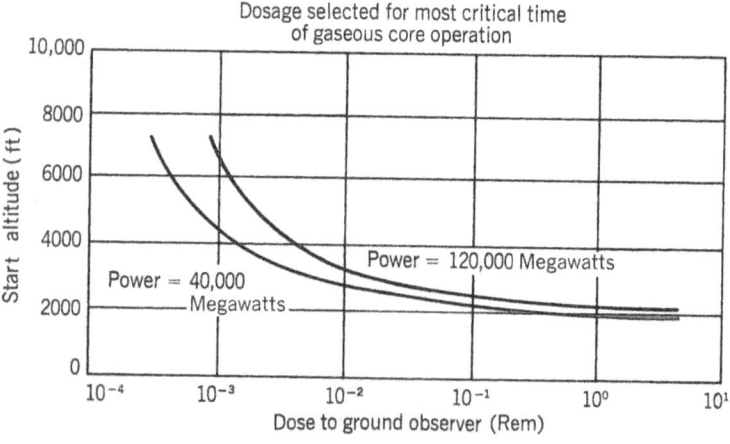

Fig. 5-18.   Dose to ground observer from gaseous core rockets.

orbital energy (Chapter 3), about 50 per cent over-all efficiency, and the fact that one kiloton of energy is defined as 3.1×10^{12} foot-pounds.) Considering that about 90,000 kiloton equivalents had been exploded in the atmosphere up to 1960, it is clear that hundreds of launches per year would be small in terms of bomb testing problems. The bomb tests prior to 1960 have been estimated to increase the natural background radiation at sea level by less than five per cent over the subsequent 30-year period.

Although spaceships present only small problems in terms of general earth atmospheric contamination, we are extremely sensitive on this subject. We wish to contaminate neither the atmosphere nor the radiation belts of any planet, nor any of space. Actually, one cannot contaminate all of space. Our own star, the sun, has been making a real effort for billions of years. The sun's energy output is equivalent to the detonation of 100 megaton bombs at the rate of 100 million per second. This occurs deep in the sun. The rest of the sun's mass is an effective shield, but the point is still the same. Every star is a nuclear event which pours debris into space far in excess of that possible from a mere spaceship. We need to be careful of local effects on planets and radiation belts, but should not confuse them with all of space.

At a specific impulse of 3000 seconds, the average exhaust velocity of a gaseous fission engine would be about 100,000 feet per second. This is not only higher than earth escape velocity; it is also higher than the 40,000 feet per second which must be added to the earth's velocity around the sun to achieve solar system escape. When operating with high specific impulse, the exhaust velocity would be several hundred thousand feet per second. For rockets of this sort, most fission products in the exhaust would be thrown completely from the solar system as long as flight programs were arranged so the exhaust jet was not pointed at either a planet or its radiation-trapping belts. Far from being a solar system hazard, such spaceships would dispose of most of their fission products automatically in a more acceptable fashion than any system limited to disposal on this planet.

Hypothetical Gaseous Fission Ship

A hypothetical gaseous fission powered ship is shown in Figure 5-19. An initial weight of one million pounds (500 tons) was selected as reasonable, although there is some doubt that a gaseous fission engine can be this small. This weight is high enough that the design payload of 100 tons is very useful, yet low enough that the ship is more compact than a modern jet transport. The cargo weights which can be carried by such ships are a large fraction of the total weight of the ship. Depending upon velocity increment, this number can be as low as 20 per cent or as high as 70 per cent of initial weight. When we lay out such a ship using water as propellant, rather than hydrogen, the design looks more like a Buck Rogers spaceship than a conventional ballistic missile. Since standard cargo densities are about 10 pounds per cubic foot, the cargo compartment takes up more space than the propellant tank. That is the way a good spaceship should be.

This particular configuration is presented as an illustration of possibilities. The lifting body configuration for atmospheric entry is idealized. Such a shape assumes it will not be necessary to supply artificial gravity for long duration missions. If artificial gravity is needed, a radically different shape, perhaps a spinning disc, might be required.

An analysis of the shielding weight for such a ship indicates 20,000 pounds to be a conservative shielding allowance. Cargo itself is about 70 per cent effective as shielding material, and most cargo would not be harmed by exposure to the radiation levels present. Some radiation is beneficial as in the case of food preservation. The expediency of never flying the ship with less than 10 per cent cargo weight, properly packaged, would effectively reduce the shielding penalty almost to zero.

Since it is desirable the ship be built of high temperature materials to permit re-entry, it is possible the same surface area can be used to radiate energy. The energy would come from the outside during re-entry and from the inside during engine operation. If the radiator area is small enough that the normal structure of the ship becomes the radiating surface,

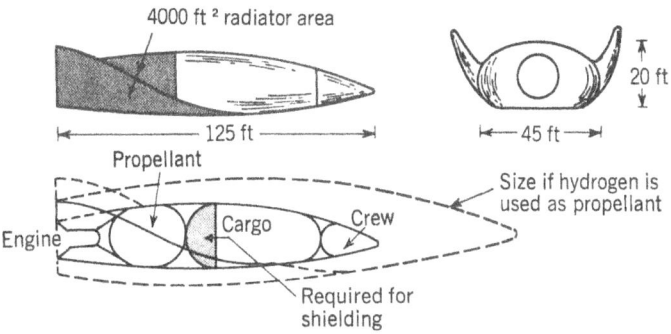

Launching weight = 1,000,000 lb (500 tons)
Cargo weight = 200,000 lb (100 tons)
Propellant weight = 650,000 lb (325 tons)

Fig. 5-19.   Gaseous fission-powered spaceship.

then the radiator weight penalties are virtually reduced to zero.

The first ships to explore the solar system will operate, for the most part, from totally unprepared facilities. Only on Earth will they operate from a spaceport. It is necessary the ships be able to take off without requiring either launch complex or take-off roll. It is also desirable for ease of cargo handling that cargo doors be close to the ground when the ship is at rest. One can spend many hours developing ship designs which meet most or all of these requirements. Many possibilities exist, just as many different ship, automobile, and airplane designs have been used. Oddly enough, in spite of extensive rocket work, there has been relatively little serious spaceship design as of 1965.

Performance curves so far have stressed the use of high specific impulses. Weight estimates assumed an engine thrust/weight ratio of 20. With the radiator area shown, the spaceship of Figure 5-19 would have an acceleration fully loaded of only 0.10 g_o at 10,000 seconds specific impulse. It could, however, operate at lower specific impulse when high thrust was required. Once orbital velocity is attained, 0.10 g_o is adequate.

About 0.20 g_0 would be required for normal take-off from the large natural satellites of the solar system. Take-off from Mars and Mercury would require reducing specific impulse to about 5000 seconds for the first 10,000 feet per second velocity. The only substantial penalty associated with low thrust/weight ratio is the necessity to climb out of Earth or Venus gravity fields with initial thrust/weight ratio of 1.25. This penalty only influences very high velocity missions with Earth launch. It can be avoided completely by one refueling from a sister ship.

High performance engines are of great use in low performance missions. Although this section has emphasized a versatile vehicle which could operate throughout the solar system, its use on lunar cargo runs is very worthwhile. Without radiators but with hydrogen propellant, the ship could deliver 100 tons cargo to the Moon 50 times a year, or 5000 tons per year. Ten such ships could deliver the same tonnage per year to the Moon which we supply to Antarctica.

If the ship were powered by an engine without radiators but operating on water, its specific impulse would be only about 1200 seconds. This means an impulsive velocity of only 38,700 feet per second. However, 30,000 feet per second is enough to attain orbital velocity, and this ship could re-enter at will. It could operate between any two points on Earth at orbital velocities carrying 20 per cent of launch weight as payload, and needing only water as propellant. The time to travel half way around Earth without exceeding 1.0 g_0 during acceleration or deceleration would be 54 minutes. (The minimum time without exceeding 1.0 g_0 is 47.5 minutes since orbital velocities beyond 36,700 feet per second require negative accelerations greater than 1.0 g_0 to hold a circular path; see Equation 3-15.)

This ship's payload fraction would be higher than a subsonic jet transport. As shown in Figure 5-9, its fuel cost with perfect containment would be only 0.3 cents per pound of propellant. This is almost a factor of 10 lower than the price of kerosene. Fuel cost to deliver cargo to 2000 miles would be less than one cent per ton-mile. To deliver cargo half way around the world, it would be one-sixth cent per ton-mile. These are very impressive numbers. Such a domesticated spaceship should have lower operating cost than any current subsonic jet or projected supersonic transport. As a commercial transport, or military cargo craft, it would be superb.

Non-Rocket Solar System Transportation

One wonders if guns could ever compete with rockets in space transportation. Once a lunar base is established, we might imagine constructing a gun accelerator several miles long for space vehicles. An electromagnetic gun would be more suitable than a gas expulsion device. The gun in space is different from on the earth's surface. Almost any velocity could theoretically be reached since there would be no air resistance either outside or inside the accelerator. Since no propellant would be thrown away, only a little more than the electrical energy equivalent of the kinetic energy of the vehicle need be expended.

It has been suggested an asteroid be brought here and placed in orbit around earth. This might be possible using nuclear bombs on the asteroid as external pulse rocket charges. Vehicles would then be fired vertically from earth with an electrical gun at minimum velocity to just reach the orbit of the asteroid. The asteroid would scoop up the vehicle, decelerating it within an electromagnetic tube and converting its kinetic energy into electrical energy. This electrical energy would be stored and used later to accelerate the vehicle to space. This represents a daring idea. It is analogous to the close flyby of a planet by a vehicle. An asteroid in earth-orbit would not have a sufficiently high gravity field to deflect a vehicle moving at 26,000 feet per second with respect to it through the almost 180 degrees required for ejection into space. Catching it in a magnetic gun, and re-accelerating it with its own energy is a way of doing this without a large gravity field.

There are two main objections to the use of electromagnetic guns. First is the length of acceleration device required. For human transportation, it is highly desirable that the acceleration be limited to about one g_o. A curve of acceleration distance versus velocity for one and two g_o's is shown in Figure 5-20. Even if solar transportation is envisioned as only Hohmann Transfers, 7800 feet per second is required for lunar escape, and the acceleration device would be 179 miles long for 1.0 g_o. Although awkward, this does not represent an impossible engineering feat. Truly convenient solar transportation will require velocities as high as 500,000 feet per second. A velocity of 500,000 feet per second, even at two g_o's acceleration, requires a 368,000 mile accelerator. Even if a magnetic gun does save rocket fuel, it is limited to low capability solar transportation systems.

One could conceive of an electromagnetic accelerator of the required length built by wrapping a 6800-mile track around the maximum diameter of the moon with electromagnetic accelerator stations at intervals. The vehicle would make 15 turns around the track if 100,000-mile acceleration distance

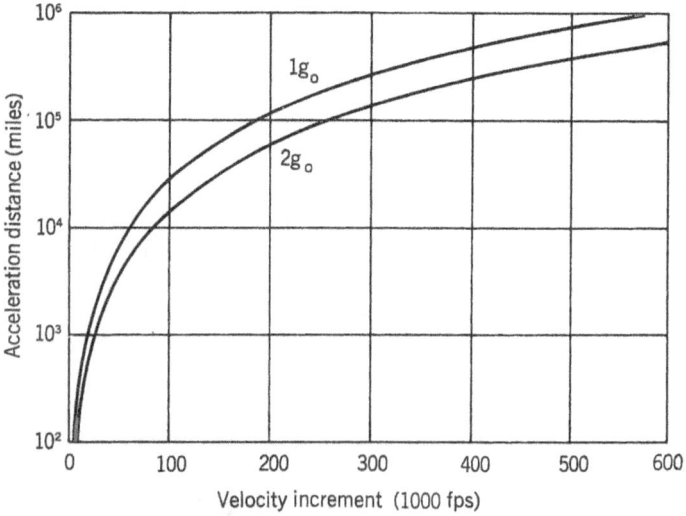

Fig. 5-20.   Acceleration distance.

were required. Compared to civil engineering feats on earth, much longer railroad tracks and greater power stations have been built, though not with the precision required for such high velocities. Even this scheme is not feasible, for now the vehicle will experience centrifugal force due to the curved track. The expression for effective gravity, Equation 3-15, shows that on the moon a velocity of only 15,000 feet per second requires one g_o and 20,000 feet per second requires two g_o's lateral force.

The other objection to electromagnetic guns is that rockets are very efficient. Figure 1-7 shows that a maximum efficiency rocket (velocity increment = 1.6 exhaust velocity) will convert 65 per cent of its exhaust energy to kinetic energy of the final weight. Including the effects of efficiency of internal energy release, a future rocket would be about 50 per cent efficient. Although the electromagnetic gun should be highly efficient, we must consider its energy source. Either chemical or nuclear energy would have to be converted to electrical energy. The best power stations on earth do this at only 30 per cent efficiency. In the case of the asteroid at earth orbit process, where initial energy comes from the vehicle's kinetic energy with respect to the asteroid, it would be necessary to store and reuse electrical energy with 50 per cent efficiency to begin to compete with rockets. Since storing of electrical energy involves converting it to other forms, the installation would almost certainly be less efficient than a properly designed rocket.

This valid point, clearly understood by Tsiolkovskiy, seems to have been missed by science fiction writers as well as engineers ever since. Rockets, properly used, are the most efficient converters of nuclear or chemical energy to useful kinetic energy yet discovered. The penalty of carrying fuel aboard is not great and is substantially less than penalties of energy conversion required in non-rocket systems. Furthermore, rockets can spread their acceleration over distances of hundreds of thousands of miles as required for passenger comfort. The gun will not rise again. *The future of space exploitation belongs strictly to the rocket. It is quite adequate for the job.*

6

Interstellar Ships
(Velocity Approaching the Velocity of Light)

Stellar Mechanics

Even in terms of the great distances and velocities already discussed, the stars are very far away. They are so far away that their distance is commonly measured with yet another unit. It is the distance light travels in one year—called a light year. Since light travels 186,000 miles per second (1000 times faster than the highest velocity in Chapter 5), a light-year is about 5.9 trillion miles, or 63,500 astronomical units. (The term parsec, an acronym for seconds of heliocentric parallax is also used. One parsec = 3.22 light years = 19 trillion miles.)

The nearest known star is Alpha (α) Centauri. It is about 4.29 light-years from the Sun. The distance from the Sun to Pluto at aphelion is 6.8 light-hours. Thus, the nearest star is about 5,500 times as far away as the most distant known planet of the solar system.

To the naked eye, the stars appear as points of light. When viewed through a telescope, many are revealed to be several stars revolving around each other. More than half the stars in the sky are multiple stars. Alpha Centauri is a triple star.

Probability of Planets

Astronomers classify stars according to their temperature and the intensity of light emitted. The rate of rotation also can be measured. Certain types of stars, known as F, G, K, and M type stars, have much lower rates of rotation than the others. This is currently considered as evidence that those stars have formed planets. Were a planetary mass created during star evolution, the star would slow down in rotation by the amount of angular momentum carried off by the planet. Ninety-seven per cent of the angular momentum of our solar system is contained in the planets, even though they constitute only 0.134 per cent of the mass of the system. We can estimate the probable number of planetary systems in our galaxy by noting the number of F, G, K, and M type stars. The suitability of planets about M type stars for life is questionable since they have low intensities and temperatures. We shall concentrate on F, G, and K types.

Alpha Centauri B is a K type star, and Alpha Centauri A, a G type, the same as our sun. It is also about the same size. Although this is a strong hint that planets have been formed there, the suitability of any such planets for life as we know it is a complicated question, since extreme temperature variations will occur on planets with highly elliptical orbits. A single star can have planets in stable orbits anywhere about it. Theoretically, however, multiple stars may only have planets in stable orbits close to one star or very far from all.

Alpha Centauri A and B revolve around each other with a period of 80.09 years on elliptical orbits with maximum separation of 47.4 astronomical units and minimum separation of 24.7 astronomical units. Unfortunately, even the restricted three-body problem has been solved for only circular orbits, and hence cannot be applied to Alpha Centauri. If Alpha Centauri A and B maintained circular orbits at the closest distance, however, planets could exist around Alpha Centauri A at distances comparable to that between Saturn and the Sun. Unless something very unusual occurs in stability effects of elliptical orbits, Alpha Centauri should be as good a source of habitable planets as a single star.

Alpha Centauri C, the nearest known star, is a flare star circling A and B at a great distance. Thus, Alpha Centauri C does come at times 0.17 light-years nearer to Earth than A or B, and accordingly, is frequently called "Proxima Centauri."

Using information on stellar types, it is possible to construct a picture of our galaxy as seen through the eyes of a starship designer. Figure 6-1 shows the 30 stars nearest to our Sun.

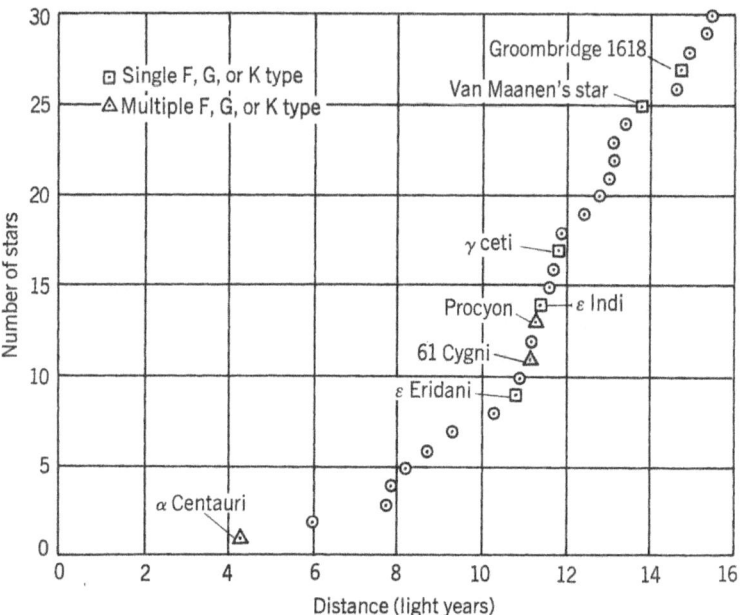

Fig. 6-1.　The near stars.

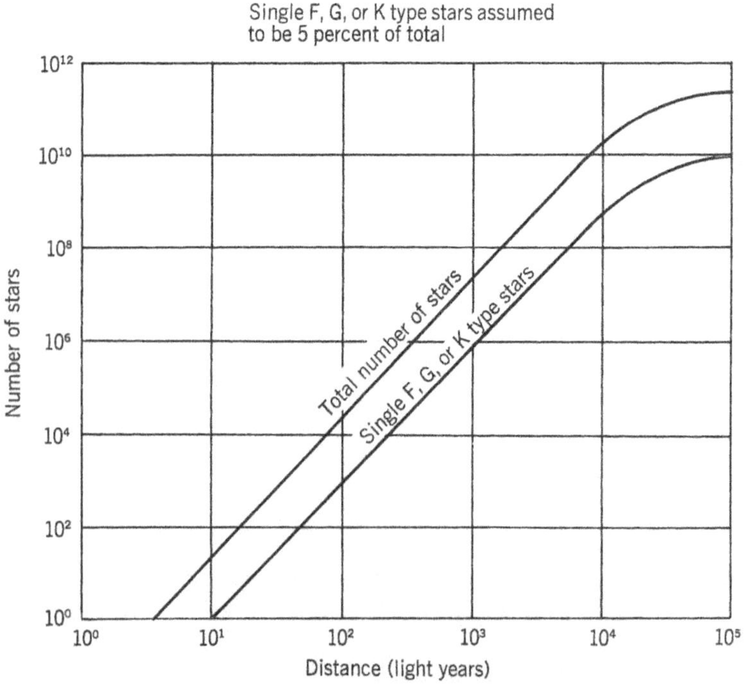

Single F, G, or K type stars assumed
to be 5 percent of total

Fig. 6-2.　The galaxy.

Multiple stars are counted as only one star. Multiple and single stars with F, G, or K type components are designated.

The nearest single stars of F, G, or K type are ϵ Eridani at 10.8 light-years, ϵ Indi at 11.4 light-years, and γ Ceti at 11.8 light-years distance. Figure 6-2 is an extension of Figure 6-1 to the whole galaxy. It assumes, based on local star counts, that single F, G, and K type stars constitute five per cent of the galactic stellar population. They may run as high as 67 per cent of multiple stars.

Communication with Intelligent Life

Project Ozma, an attempt to detect intelligent signals from stars, was made in 1960. The radio telescope antenna was directed toward γ Ceti and ϵ Eridani. Neither α Centauri nor ϵ Indi were visible at the antenna location. The attempt was unsuccessful. The probability of there being transmissions from an alien race, however, remains of interest. To say the least, it is difficult to estimate this probability. We have reason to believe (due to research into chemical evolution during the past two decades) that life would arise spontaneously on any planet with temperatures not unlike those of the earth. There remains still the question of rise of intelligence and culture. Furthermore, if a culture reaches the point where it wants to communicate, how long will it have this urge? Perhaps after another 5,000 years, the human race will not have a scientific culture. It may be something entirely different. Our descendants may not care about communicating with anyone.

Figure 6-3 adds an estimate of the occurrence of intelligence throughout the galaxy to Figure 6-2. The bottom curve, labelled Social Interest, assumes that life would develop at each single F, G, and K type star; that after five billion years, it would produce a society; and that the average society would be actively interested in communicating with other civilizations for only about 50,000 years. Astronomical and geological evidence agree that the planet Earth was formed about five billion years ago. Thus, our star required five billion years to produce a society. Races on other planets could have evolved

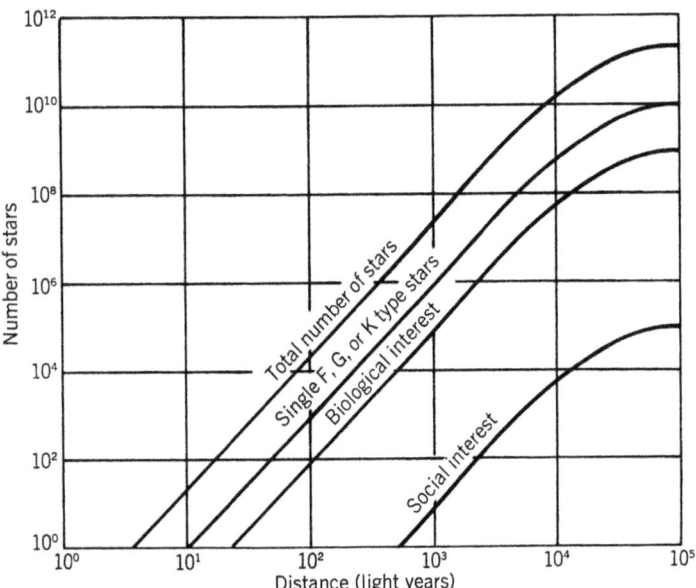

Fig. 6-3.   Hypothetical galactic community.

slower or faster. Unfortunately, we only have one data point on this subject. We do not even have one data point on the 50,000-year active communication assumption. Speculation ranges all the way from only a few decades—if scientific cultures tend to destroy themselves—to millions of years—if they learn to live in peace after releasing nuclear energy. It is not surprising, in view of the great distance to the nearest probable communicating civilization, that many people dismiss going to the stars. If it is only societies which are currently communicating who are of interest, then perhaps it is appropriate to just listen from earth and hope to learn that way.

There is, however, another class of stellar system which could be of interest. A spaceship that travels to the stars is, in a way, a time machine. Deliberate communication records of human society on this planet cover thousands of years. Ancient races have communicated to, if not with, us by their writings. We have been able to look hundreds of millions of years into the past, learning things of biological interest, such as patterns of the development of life. If one goes to the planets of a distant star with suitable scientific personnel and equipment, one can look both back and ahead in time compared to the limited real time contact with any currently communicating society. If we assume 500 million years as the time during which a planet is of biological interest, based on our use of data from a comparable time span on this planet, the curve labelled Biological Interest on Figure 6-3 results.

Acceleration

The great distances to the stars make it clear that no interstellar travel, even to the nearer stars, can occur within reasonable travel times in terms of a human life unless an appreciable amount of the velocity of light can be attained. If a constant acceleration of one g_o could be maintained, it would take one year to reach the velocity of light. Consequently, accelerations of this order are mandatory for starships. There appears to be no place for low-thrust electrical rockets requiring millenia to generate any significant portion of the velocity of light in interstellar travel.

A starship starting from earth and accelerating at a constant one g_o would reach 32 astronomical units in 11.4 days and would be going at 3.24 per cent of the velocity of light. A tracking network covering the entire solar system would observe only the beginning of such a flight.

Time Dilation

If ships can approach the velocity of light then the various effects predicted by Einstein's Theory of Relativity must be considered. Of major consequence is the limitation that the velocity of material particles cannot exceed that of light. Since our galaxy is about 100,000 light-years across, this restriction seriously limits interstellar travel within our galaxy, let alone travel to other galaxies.

In addition to the limitation on maximum velocity, the

Theory of Relativity predicts that time dilation will occur as the ship approaches the velocity of light. That is to say, time, as measured in the ship, will appear to go slower than time measured by those who remain on earth.

The relation between earth time and ship time is given by:

$$t_S = t_{EA}\sqrt{1 - (V/c)^2} \qquad (6\text{-}1)$$

where t_s = ship time; t_{EA} = earth time; V = ship velocity, and c = velocity of light. Figure 6-4 shows earth time and ship time versus ship velocity. The times are shown compared to earth time if the ship were travelling at the velocity of light. One must approach quite closely the velocity of light to get any appreciable time dilation effect.

Attainment of very high velocities might be considered the physicists approach to time dilation. There are other disci-

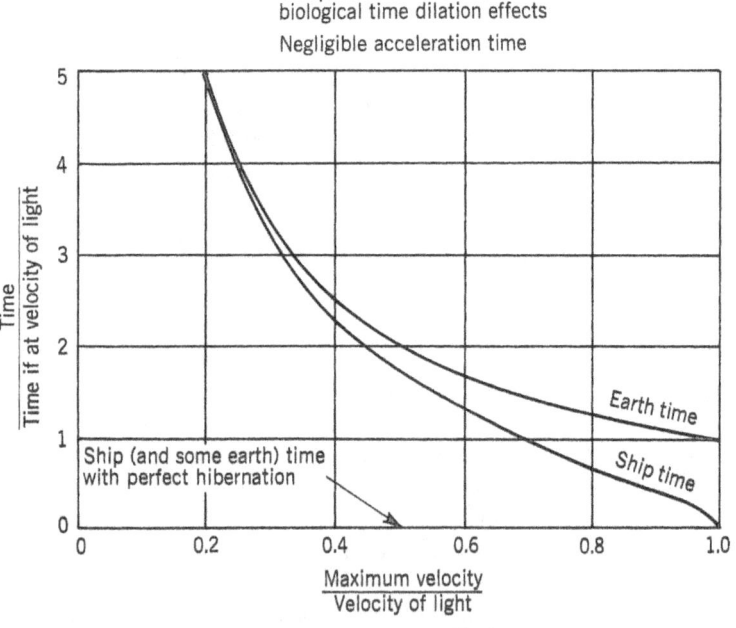

Comparison of relativistic and
biological time dilation effects

Negligible acceleration time

Fig. 6-4.   Interstellar travel time dilation effects.

plines, however. One is biology. Some fascinating things have been happening in this field recently. Biologists seem to be getting closer, by deep-freeze and other techniques, to learning the secrets of hibernation—biological time dilation. If achieved, it could be applied to the crew of a starship, making extremely high velocities unnecessary. At one-third the velocity of light, the elapsed time is only three times as great in earth years as it would be if the ship travelled at the velocity of light. Physical time dilation could only be applied to those in the starship. Biological time dilation could be applied both to them and to personnel on earth. The latter prospect has profound implications quite independent of space travel.

The question, then, of whether starships must travel at almost the velocity of light or only about one-third that velocity depends greatly on a different discipline from physics. If biologists succeed in accomplishing true hibernation, they will exert a greater leverage on starship design than those who would attempt to design starships capable of almost the velocity of light. Many analyses have been made with velocities about one per cent lower than the velocity of light. The conclusions are that the ships required would be ridiculous, and, therefore, interstellar travel even to the nearer stars is ridiculous. This represents a very naive systems analysis of interstellar travel.

Escape from the Galaxy

As Figure 6-2 indicates, our galaxy is so large (100,000 light years across), and contains so many stars (100 billion), that it would seem large enough to occupy us quite a while. This does not prevent consideration of the question of escape from the galaxy. The galaxy is rotating and an escape velocity is required, as for other gravity fields. Not all of the galactic mass (estimated to be equivalent to 200 billion suns), however, can be assumed to be concentrated at a central point. Indeed, some lie outside of the sun's orbit. Consequently, a somewhat different analysis of escape velocity is required, but will not be presented here.

The sun is estimated to be on a roughly circular orbit, 30,000 light-years from the galactic center with a period of about 200 million years. It is moving at about 720,000 feet per second, and galactic escape velocity at this location is about one million feet per second. Hence, a solar hyperbolic excess velocity of about 300,000 feet per second is required for galactic escape. This would have to be at a somewhat steep angle to the ecliptic, however, since the earth's orbital plane is inclined at about 60 degrees to the plane of the galaxy.

Fusion Rockets

Examination of the propulsion requirements for interstellar travel naturally leads to a discussion of fusion rockets. The ratio of initial to final weight of an undiluted fusion rocket, as a function of the ratio of maximum velocity to velocity of light, is shown in Figure 6-5. The weight ratio is that required both to accelerate to the velocity shown and to decelerate to zero on arrival. When relativistic effects are included in the derivation, the basic rocket equation becomes:

$$\frac{w_I}{w_F} = \left[\frac{1 + \dfrac{\Delta V}{c}}{1 - \dfrac{\Delta V}{c}} \right]^{c/2v_{ef}} \tag{6-2}$$

For the case of rocket braking on arrival, plotted in Figure 6-5, this expression becomes:

$$\frac{w_I}{w_F} = \left[\frac{1 + \dfrac{\Delta V}{c}}{1 - \dfrac{\Delta V}{c}} \right]^{c/v_{ef}} \tag{6-3}$$

Although Equation 6-2 does not look like Equation 1-17, they are the same for small values of $\Delta V/c$. A curve calculated by ignoring the relativistic terms is included in Figure 6-5 for comparison.

The gaseous fission rockets of the previous chapter would seem to be unsuited for interstellar ships with velocities of

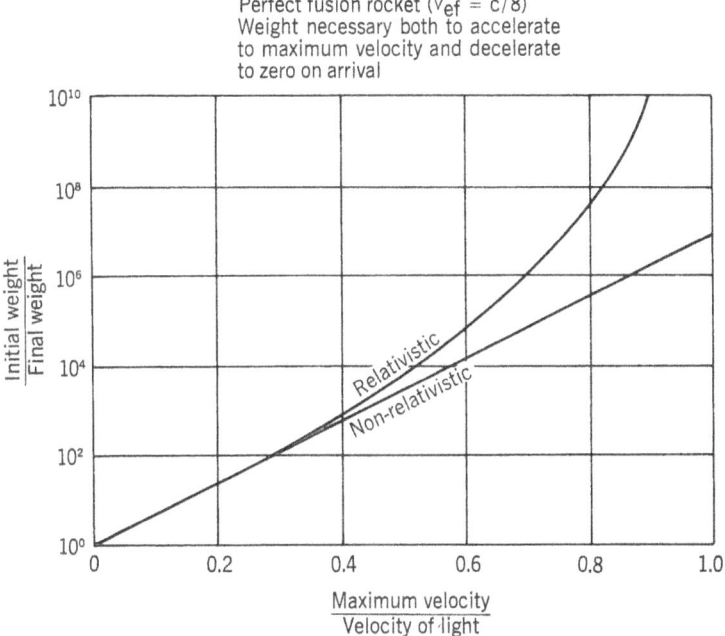

Fig. 6-5.   Fusion starship weight ratio.

the order of one-third the velocity of light. The undiluted specific impulse of fission reactions is about 0.3 that of fusion reactions. Equation 6-3 shows that a weight ratio of 2.5 million would be required for a fission rocket (less than 100 for a fusion rocket) to achieve even 0.25 the velocity of light. The weight of the earth would be required to drive one pound to 76 per cent of the velocity of light with undiluted fission and 99.8 per cent with undiluted fusion.

The same characteristics which cause fusion power to be of extreme interest on earth would also be a great help in interstellar travel. Fission fuel is expensive—around $5,000 per pound. The fuel cost of a fission ship with initial to final weight of one million would be five billion dollars per pound of final weight. There is hope that fusion reactors can be made to operate on cheap fuels. Heavy water today costs less than

$25.00 per pound. Deuterium (heavy hydrogen), a possible fusion fuel, costs about five times that. If a fusion rocket of initial to final weight of 100 were used, the deuterium would cost $10,000 per pound of final weight. Since fusion reactions release about 10 times the energy per pound of fission reactions, fusion fuel at $30 per pound is about 1600 times better than fission fuel in energy per unit cost. A starship with maximum velocity of 0.3 the velocity of light, however, would generate a total kinetic energy 350,000 times that of a one million foot per second solar-system spaceship. Starship fuel costs are high, even assuming fusion. For them to be as economical as solar-system spaceships, fusion fuel would have to be the same price as hydrogen-oxygen.

In addition to cost, it is likely that fusion rockets would require less shielding than fission rockets. Fusion reactions do not directly create fission products to complicate shielding. Although neutrons will likely be produced so that shielding will be necessary, the amount and variety of particles and radiation produced by fusion reactions are not as bad as those produced by fission reactions, as long as reasonable care is used in selecting materials subject to neutron bombardment.

Fusion rockets will have cooling problems, just as the gaseous fission rockets discussed in Chapter 5. The power generated by a typical small starship is shown in Figure 6-6. This is the initial power given by Equation 1-11, assuming 100 per cent efficiency and an initial thrust/weight ratio of one. This is a small ship with final weight of only 10,000 pounds. The payload weight will be even less, due to shielding requirements. Yet, if designed for a maximum velocity of 30 per cent that of light, this ship would initially generate about 1000 times the power of a gaseous fission engine of one million pounds thrust.

The control and re-radiation of superfluous energy without vaporizing the ship is, at the moment, a totally unsolved problem. The magnitude of such energy is uncertain, but some estimates are as high as 20 per cent of the total energy. Since fusion reactors require no moderators, however, perhaps most of this energy can be permitted to escape directly. The problem can be greatly alleviated by engine configurations utilizing materials highly transparent to the radiations, with payload and propellant shielding placed only where necessary. This is a starship analogy of the ordinary chemical rocket technique of controlling large energy by avoiding most of it. As a matter of interest, even with the small final weight in Figure 6-6, such a ship would initially have to generate power equivalent to the total power output of the sun, were it to generate 97 per cent of the velocity of light.

It must be emphasized that, at the moment, we seem to be far away from fusion rockets. Fusion plasmas have not yet been stabilized well enough to contain a self-sustaining fusion reaction. All of the systems under test for controlled fusion

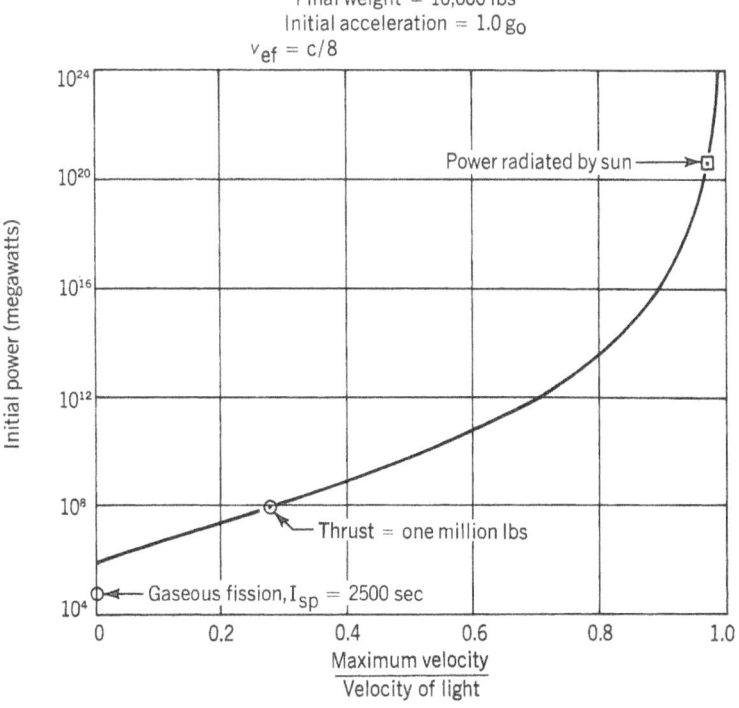

Fig. 6-6.   Fusion starship power.

power use magnetic containment systems, since it seems fundamental that any contact with solid materials would quench a fusion reaction. Even with the most modern cryogenic magnets, the thrust/weight ratio of fusion rockets would be less than 0.01. Hundreds of years would be required to accelerate to an appreciable portion of the velocity of light.

Should fusion rockets ever be developed, they, like almost all high performance engines, could be helpful in lower performance missions. Figure 6-7 compares fuel costs of both fission and fusion rockets as a function of specific impulse. There have been suggestions that fusion rockets be made to operate with deuterium and helium-3 because this reaction would produce very few neutrons and thereby decrease shielding problems. But helium-3 is essentially as expensive as fission fuel and the gain of fusion over fission would not be great.

If a fusion rocket operated on fuel which cost only $30 per pound, the resulting effect on solar system transportation would be spectacular. A specific impulse of 500,000 seconds

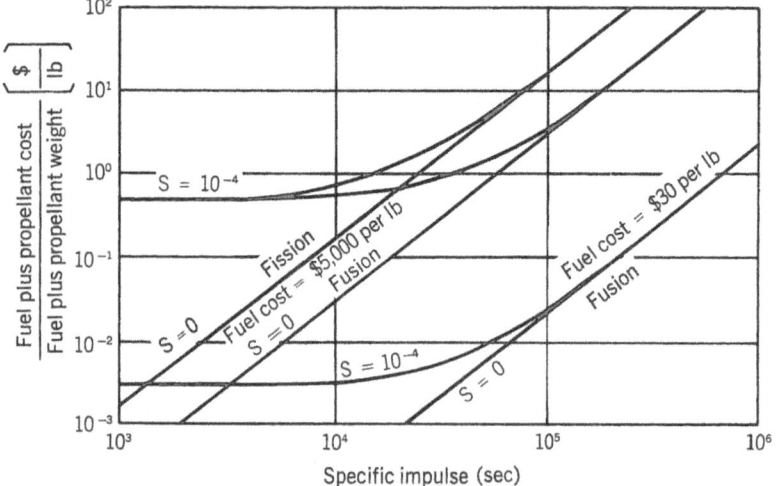

Fig. 6-7.   Cost of nuclear rocket fuel and propellant.

would require a fuel plus propellant cost of about 50 cents per pound. A single stage ship with such an engine could generate about 16 million feet per second with 20 per cent payload. (Sixteen million feet per second is about 3000 miles per second. Such a ship, passing the Earth at high velocity, would cross the United States in one second.) Even using one-half this velocity for braking, and limiting the ship acceleration to one g_o such a ship could cover the distance from the Sun to Pluto in 25 days (approximately 40 per cent of the time accelerating and decelerating). Thus, fusion rockets, if ever built with high thrust/weight ratios to operate on cheap fuels, will be vastly better than gaseous fission engines for solar system transportation.

We can consider an even more routine use of fusion rockets. Just as domesticated gaseous fission spaceships could make excellent terrestrial transports, we can examine the use of starship engines in terrestrial transports. In this case, the ability of gaseous fission ships to generate the highest usable velocities, considering passenger comfort, at quite low fuel costs, does not leave much room for improvement. Certainly, a fusion rocket with perfect containment would lower fuel and propellant costs by a factor of 1000 under those of a perfect containment gaseous fission system and would solve the fission product disposal problem. By that time, however, gaseous fission systems may already have reduced fuel cost to a minor portion of total operating cost. Fuel cost of the fusion rocket would be lower even than current fuel cost of water transportation, and the possibility of a revolutionary transportation system, combining the low cost of current water transport with the short travel time of orbital transport, cannot be discounted.

Photon Rockets

A photon rocket is one which emits only radiation as exhaust. Almost everyone has held a photon rocket, a flashlight, in his hand. The problem is not in creating a photon beam, but in creating an intense beam with high thrust and long duration. If the energy source is a fission or fusion reactor, only a fraction of total mass can be converted to radiation. If the reaction products are ejected from the ship at zero relative velocity, then the photon rocket behaves as if it had a fictitious exhaust velocity given by:

$$v_{ef} = \epsilon c \qquad (6\text{-}4)$$

where ϵ = fraction of mass converted to radiation. Equation 6-4 put in the relativistic rocket Equation gives:

$$\frac{w_I}{w_F} = \left[\frac{1 + \dfrac{\Delta V}{c}}{1 - \dfrac{\Delta V}{c}} \right]^{\frac{1}{2}\epsilon} \qquad (6\text{-}5)$$

Comparing Equations 6-5 and 6-2 for fusion reactions shows that photon rockets are much less efficient than pure fusion rockets. Rejecting the reaction products at zero velocity, even though the beam has the specific impulse of light (30.5 million seconds), is not as efficient as distributing the energy into all particles exhausted. A fusion photon rocket, however, would be more efficient than a diluted gaseous fission engine.

At the moment, the possibility of generating photon beams of any substantial thrust seems remote. The exhaust power of a photon beam is given by:

$$P_{ef} = Tc \qquad (6\text{-}6)$$

This is twice the value one would expect from Equation 1-11. Using Equation 6-6, a power of 1,333 megawatts is required per pound of thrust. To radiate this power from one square foot would require a temperature of 40,000° F. (see Equation 5-7). At a radiating temperature of 200,000° F., the power equivalent of 600 pounds of thrust would be generated by one square foot of area. These powers are seriously being considered for gaseous fission engines, but the energy flux in that case is poured directly into the propellant. We do not know how to build a transparent reactor to let out this intense light. To prevent vaporization, a mirror reflecting such a beam might have to be so far away that its size would be prohibitive. Photon rockets are not in near sight, but in the heart of gaseous fission engines there may be created light sources so intense that, if unleased, they would produce respectable photon rockets.

Mass Annihilation Rockets

All nuclear rockets are a form of mass annihilation rocket since the energy release comes from the conversion of mass to energy. By a mass annihilation rocket, we specifically mean one which converts all its fuel mass to energy, not the partial conversion of normal nuclear reactions.

Total mass annihilation is theoretically conceivable and has been observed in particles. Antiparticles have been found for all known particles. When a particle and its antiparticle combine, they annihilate each other with release of the total energy equivalent of their masses. Antideuterons (antihydrogen nuclei) have been observed. Someday we presumably will be able to manufacture antimatter with antiparticles in the nucleus and positrons around the nucleus, as electrons surround the nuclei of normal atoms. It is conceivable that we will discover stars and planets made of antimatter but no indication of this possibility has yet occurred.

Even if we could create antimatter at will, its production and use present fantastic engineering problems. Any contact with normal matter must be avoided, and magnetic storage in very hard vacuums must be arranged. It would not be surprising should we find that antimatter could only be produced in quantity and stored in space. The very good, almost unlimited vacuum of space may never be reproducible on a planet. Even to handle antimatter in space would be a very tricky operation with a spectacular intolerance to minor engineering mistakes.

The most effective use of mass annihilation may be in normal fusion reactions. From a propulsion point of view, a pure fusion rocket is simply a mass annihilation rocket which operates at a dilution ratio of about 100 due to the non-annihilated mass carried along. Operation at this ratio solves all the manufacturing and handling problems mentioned, and one

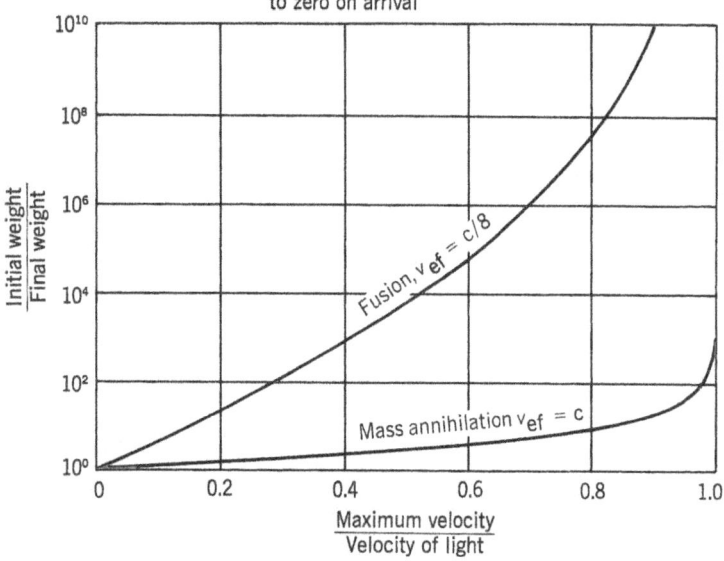

Fig. 6-8.   Starship weight ratio.

should be certain that the gains in using an undiluted mass-annihilation rocket are worth the effort.

With this type of rocket, nothing is ejected from the rocket but radiation. They are the ultimate in photon rockets. The rocket equation for mass annihilation rockets is obtained by placing $v_{ef} = c$ in Equation 6-2 (or $\epsilon = 1$ in Equation 6-5) to give:

$$\frac{w_I}{w_F} = \left[\frac{1 + \dfrac{\Delta V}{c}}{1 - \dfrac{\Delta V}{c}}\right]^{\frac{1}{2}} \qquad (6\text{-}7)$$

For the case of rocket braking, this expression becomes:

$$\frac{w_I}{w_F} = \left[\frac{1 + \dfrac{\Delta V}{c}}{1 - \dfrac{\Delta V}{c}}\right] \qquad (6\text{-}8)$$

A comparison of final to initial weight ratio for fusion and

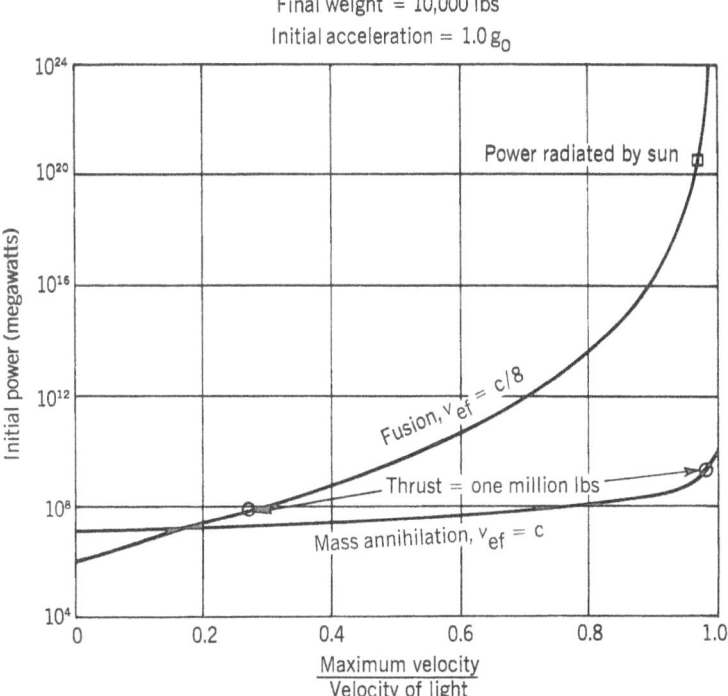

Fig. 6-9.   Starship power.

mass annihilation rockets is shown in Figure 6-8. Mass annihilation rockets would be substantially lower in weight for a given mission than fusion rockets, but even they would not permit reasonable sized ships if maximum velocity were much beyond 95 per cent of the velocity of light. Figure 6-9 shows the initial power generated by mass annihilation rockets compared to fusion rockets for equal final weight. At 30 per cent of the velocity of light, the mass annihilation rocket requires only 20 per cent of the initial power of a fusion rocket. It would generate over 80 per cent of the velocity of light with the same power as a fusion rocket at 30 per cent of light velocity.

Pure mass annihilation rockets with high thrust/weight

ratios would have fantastic performance. They represent an interesting end point in the utilization of mass and energy, and would obviously be examined thoroughly if any hope of producing and controlling antimatter were to arise.

Interesting approximations to the rocket equation are possible for the case where a very close approach to the velocity of light is contemplated with a mass annihilation rocket. The time dilation expression, Equation 6-1, becomes:

$$t_S = t_{EA} \sqrt{\left(1 + \frac{V}{c}\right)\left(1 - \frac{V}{c}\right)} = t_{EA} \sqrt{2\left(1 - \frac{V}{c}\right)} \tag{6-9}$$

Likewise, Equation 6-8 becomes:

$$\frac{w_I}{w_F} = \left[\frac{2}{1 - \dfrac{\Delta V}{c}}\right] \tag{6-10}$$

If one considers the acceleration time to be a negligible portion of total travel time, then Equation 6-9 may be used to approximate the time dilation effect over the whole journey rather than representing only cruising velocity. Then V in Equation 6-9 can be replaced by ΔV, and Equations 6-9 and 6-10 combined to give:

$$\frac{w_I}{w_F} = \left(\frac{2t_{EA}}{t_S}\right)^2 \tag{6-11}$$

This relation between time dilation achieved and rocket weight may be used to approximate extreme missions. If one attempted to travel to the Andromeda nebulae within a human crew lifetime, a ship time of 20 years (one way) would have to be achieved while travelling 2 million light-years. Time would have to be dilated by a factor of 100,000, and the initial to final weight of ship would be 4×10^{10}. Even then it would have to be refueled at Andromeda for the return journey.

The initial power can be estimated easily using Equation 6-6. If our hypothetical intergalactic ship had one million pounds final weight, it would have to generate 5.33×10^{19} megawatts—or about 16 per cent of the power output of the sun.

Even if we envisioned such tremendous ships, it is not clear they can be driven to such velocities. In rocket performance, space is normally assumed to be a vacuum. Even intergalactic space is not, and though only a few particles exist per cubic foot, at the velocities just discussed each hits the ship with great energy. Not only will the ship experience drag, but the energies are so high that the ship may be destroyed. According to relativistic mechanics, the kinetic energy per unit mass of a particle at 86.6 per cent of the velocity of light is the same as the energy release of total mass annihilation. Even the vacuum of space becomes lethally destructive as the velocity of light is approached. If one travelled in regions where only ionized particles existed, it is possible that they could be deflected magnetically.

The Andromeda ship would return after 4 million years had passed on earth. Were a mass annihilation rocket with maximum velocity of 0.8 the velocity of light used, it would return in 5 million years. The 20 per cent longer travel time reduces the initial power to about 10^{10} megawatts—less than one-billionth of the previous ship. It seems clear that biological time dilation is the key to deep interstellar travel, with the maximum velocity of ships set by the lethal effect of interstellar matter.

Chapter 6 closes with the realization that *interstellar travel to the nearer stars is not ridiculous, as many people think*. Travel times might well be measured in decades, and fusion-powered rockets, which we do not yet know how to build, would be required. Travel deep in the galaxy or to other galaxies, however, seems remote. With excellent biological time dilation, it could be done, but time passage on earth would be measured in hundreds of thousands or millions of years.

7

Outlook

The previous chapters in this book covered a very wide spectrum of propulsion capabilities. Starting with mere flying arrows, we have traced a path up through today's fairly impressive engines of destruction to plans for tomorrow's monster rockets which will first place man on different planets. Beyond that, plausible spaceship designs, capable of turning the entire solar system into a human backyard, just as transport airplanes turned the once formidable Atlantic and Pacific oceans into little more than duck ponds, have been discussed. Still further, the dim outline of powerful starships can be seen if one knows where to look in the hazy forest of technical progress.

Several basic questions come to mind as a result of the awesome possibilities presented by the thrust of the human race into space. They usually boil down to two. Why go to space? And—what do we do when we get there?

The first question, I believe, is the wrong question to ask. Obviously, the human race is going to space. The history of man predicts this—wherever it has been possible to go, humans have gone. Their reasons have not always been logical, and most of the time, they've been wrong. Christopher Columbus set out to solve the problem of trade between Spain and China. Today, the problem is still unsolved. Whether or not we have good reasons, we are on our way. Thoughtful, intellectual people will question and decry the large expenditure but it was ever thus. Not a shred of evidence exists, however, to show that the thoughtful people of today can predict the future any more accurately than those of the past. Unless history is completely wrong, mankind will proceed into space.

The United States may or may not be the leader in this effort, but that is another question.

If one must ask a question related to the why of our going to space, it seems to me the proper question should be, why has the human race always done what it has done? Why did we explore the New World? Why did we waste the money buying Manhattan Island from the Indians? Why did we build airplanes when everyone knew railroad trains were fast enough? What is it about the restless drive of the human race which makes it always do these things? These are questions worth considering, but the question of why humans are going to space is, of itself, meaningless if considered apart from the other questions. Modern students of philosophy have pointed out that philosophical progress almost always develops when people realize that, not only do they not have the right answers to certain questions, but that they have also been asking the wrong questions. Perhaps this is true regarding man's thrust into space.

The second question—"What do we do when we get there?"—is heavily colored by the technical possibilities outlined in this book. We can clearly foresee the placing of a few men on the Moon and also on such near planets as Mars. Beyond that, vast technical controversies exist as to what future possibilities in space may be. This book makes clear the point that there is no fundamental reason why we should not be able to open up the whole solar system for human transportation, and pay roughly the same price for such a transport system as we have become accustomed to paying on this planet. The feasibility of such economical space transportation greatly colors our future in space.

Unless we falter, we will produce spaceships and we will expand the basic operations of the human race throughout the entire solar system. This could well create a new Renaissance affecting all areas of human life, not merely the rocket engineers or space scientists. Geologists, after examining various planets in the solar system, will undoubtedly find vast storehouses of new materials and knowledge. The same will be true for biological scientists. It is easily conceivable that one day the planet Mars will be the subject of most books on botany.

Space is the one place where we can obtain natural resources without damaging either the earth's ecological balance or its natural beauty. It is easily conceivable that some day specialized refining industries, using processes based on the almost unlimited vacuum of space, will spring up, perhaps in the asteroid belt as well as on planets or natural satellites. The concept of developing space resources in the future is no more a dream or less scientifically noble, than attempts to create terrestrial power from fusion reactions or to move into the sea for food supply purposes.

The cultural aspects of life will be greatly enriched out there. No terrestrial artist has seen a sunset on Mars or watched Saturn and its rings set from the vantage point of a moon such as Titan. The plunge of high performance spaceships deep into unknown territories, where even their tremendous control of energy may not always be adequate, should supply a new basis for drama, music and art far beyond mere terrestrial achievements. Some day there may even be a cathedral on the Moon. At one-sixth gravity and with no atmospheric wind loads as design conditions, a lunar cathedral should be far grander than anything possible to construct on Earth, even if eroded somewhat by meteoroids. In fact, there are no limits to the side effects. Rather, they will be the primary heritage in the centuries to come, creating a new dimension of emotional, spiritual, social, and mental stimulants beyond anything man has yet experienced.

The outlook, then, is for an exploitation of the solar system in typical human fashion. It may take decades or centuries. It will not start, though, until enough of our currently awkward first steps into space have been achieved so that we overcome our excitement and settle down to the hard-headed engineering of spaceships. It should be clearly understood that it is only such engineering which stands between us and the convenient exploration of the solar system. No new scientific discoveries are required because, basically, the price of the energy involved is reasonable.

The stars are a different matter, though. We can foresee ways of getting to the nearer stars without invoking grotesque human experiences such as 1000-year travel times. Even such times may some day appear normal if hibernation techniques become common on earth and in space. But more than mere engineering cleverness is involved in building the propulsion equipment. We do not know how to handle the energies involved. We may well be as close to starships as Goddard was to Sputnik, but it is hard to predict the rate of progress.

Clearly, the only real limitation on the future, as is foreseeable in this book, is Einstein's apparent proof that we will not be able to drive starships faster than the velocity of light. History may repeat itself, however. One can dream that, when the first sub-light starships are prepared for their voyages from a base system which covers the entire solar system, perhaps the most intriguing question among the engineers and scientists of that decade will be the desirability of an assault on the light barrier. It is not prudent, however, to ask this question today; it is not considered the right question.

List of Symbols

a	Semi-major axis of orbit	KE	Kinetic energy
a	Acceleration (Equations 1-2, 2-9, 5-6	KE_o	Kinetic energy if initial velocity were zero
A	Area	L/D	Ratio of lift force to drag force
A.U.	Astronomical unit (92.96 million miles	m	Mass
c	Velocity of light (983 million fps)	M	Molecular weight
		p	Pressure
D	Drag force	p_{at}	Local atmospheric pressure
E	Energy		
E_A	Airplane energy	P	Power
F	Force	P_{pu}	Pump power
f	Fraction of nuclear energy release which is thermally effective	P_r	Power radiated
		P	Orbital period (Equations 3-3, 3-4, 3-13, 4-4)
g	Acceleration of gravity		
g_{ef}	Effective acceleration of gravity	P_s	Synodic period
		PE	Potential energy
g_o	Standard acceleration of gravity (32.174 ft/sec^2)	PE_{MAX}	Maximum potential energy of gravitational field
h	Altitude		
h	Enthalpy (heat content) per unit weight (Equation 1-12)	r	Radius from center of gravitational field
I_{sp}	Specific impulse	r	Radius of sphere (Equations 4-7 and 4-8)
I_{sps}	Specific impulse of propellant at temperature of solid material	R	Standard radius of Earth corresponding to g_o (20.86 million feet)
J	Mechanical equivalent of heat energy (778 ft-lb/BTU)	s	Horizontal range
		t	Wall thickness
k	Ratio of specific heat of a gas at constant pressure to that at constant volume	t_b	Power plant operating time
		t_{EA}	Earth time
		t_f	Time of flight

213

t_s	Ship time
T	Thrust force
T	Absolute temperature (Equations 1-13, 1-14, and 5-7)
v_e	Exhaust velocity
V	Flight velocity
V_{c_o}	Circular velocity at R
V_{ch}	Characteristic velocity
V_∞	Hyperbolic excess velocity
w	Weight
w_{pr}	Weight of propellant
w_α	Weight of power supply
$\dot{w}$	Propellant flow rate
α	Power plant specific weight
γ	Flight path angle
δ	Ratio of structural to propellant weight
Δp	Pressure rise
ΔV	Velocity increment
ΔV_D	Drag loss
ΔV_g	Gravity loss
ΔV_{EN}	Velocity equivalent of energy expenditure
ϵ	Nozzle area ratio
ϵ	Fraction of mass converted to radiation (Equations 6-4, 6-5)
λ'	Ratio of propellant to total propulsion system weight
ρ	Fluid density
σ	Material stress

Subscripts

a	Apogee
ac	Acceleration
B	Bullet
c	Circular
co	Combustion chamber
e	Nozzle exit
E	Escape
ef	Effective exhaust
F	Final
G	Gun
i	Inner
I	Initial
o	Outer
p	Perigee
P	Planet
t	Nozzle throat
UL	Useful load
V	Vehicle

Astronomical Symbols

☉	Sun
☿	Mercury
♀	Venus
⊕	Earth
♂	Mars
♃	Jupiter
♄	Saturn
♅	Uranus
♆	Neptune
♇	Pluto

Bibliography

Buchheim, Robert W. (editor), *Space Handbook*. New York: Random House (Modern Library Paperbacks), 1959.

Cameron, A. G. W. (editor), *Interstellar Communication*. New York: W. A. Benjamin, Inc., 1963.

Deutsch, Armin J. and Wolfgang B. Klemperer (editors), *Space Age Astronomy*. New York: Academic Press, Inc., 1962.

Dornberger, Walter, *V–2*. New York: The Viking Press, 1954.

Gibney, Frank B. and George J. Feldman, *The Reluctant Space-Farers*. New York: The New American Library, 1965.

Goddard, Robert H., "A Method of Reaching Extreme Altitudes," *Smithsonian Miscellaneous Collections, LXXI*, Washington, D.C., 1919.

Goddard, Robert H., "Liquid Propellant Rocket Development," *Smithsonian Miscellaneous Collections, XCV*, Washington, D.C., 1936.

Goodwin, Harold Leland, *Space: Frontier Unlimited*. New York: D. Van Nostrand Co., 1962.

Hohmann, Walter, *The Attainability of Heavenly Bodies*. Munich-Berlin, 1925 Technical Translation F–44, Washington, D.C., National Aeronautics and Space Administration, 1960.

Langer, Susanne K., *Philosophy in a New Key*. New York: Mentor, The New American Library, 1948.

LeGalley, Donald P. and John W. McKee (editors), *Space Exploration*. New York: McGraw-Hill Book Company, 1964.

Ley, Willy, *Rockets, Missiles and Space Travel*. New York: The Viking Press, 1961.

Sagan, Carl, "Direct Contact Among Galactric Civilizations by Relativistic Interstellar Spaceflight," *Planetary and Space Science*, Volume II, 1963.

Seifert, Howard (editor), *Space Technology*. New York: John Wiley and Sons, 1959.

Sullivan, Walter, *We Are Not Alone*. New York: McGraw-Hill Book Company, 1964.

Sutton, George P., *Rocket Propulsion Elements*. New York: John Wiley and Sons, 1963.

Tsiolkovskiy, K. E., *Exploration of the Universe with Reaction Flying Machines*, Russia, 1903. Technical translation, F–237, *Collected Works of K. E. Tsiolkovskiy*, Washington, D.C., National Aeronautics and Space Administration, 1965.

Glossary

Ablating material. Material which dissipates heat by vaporizing or melting.

Absolute temperature. Temperature value relative to absolute zero.

Acceleration. Rate of change of velocity.

Angular momentum. That momentum which causes rotation of a trajectory.

Antimatter. Matter with nucleus of antiparticles surrounded by positrons.

Antiparticle. Particle with opposite electrical charge from normal matter, e.g. the antiproton has a negative charge compared to the proton's positive charge.

Aphelion. Point on the orbit of an object about the sun which is farthest from the sun.

Apoapsis. Point on the orbit of an object which is farthest from the source of gravitation.

Apogee. Point on the orbit of an object about the earth which is farthest from the earth.

Arrival window. Time during which a celestial body is so located that it is possible to arrive at it for a given velocity expenditure.

Astronomical unit. Unit of length defined as the average distance between earth and sun, about 92.96 million miles.

Ballistic missile. Missile without wings which achieves its range by generating velocity and then following a ballistic path (like a gun) to the target.

Black body. A "body" which absorbs all of the electro-magnetic radiation striking it. One which neither reflects nor transmits any of the incident radiation.

British thermal unit (BTU). Amount of heat required to raise one pound of water 1° F at 60° F. Equal to 252 gram-calories.

Capture orbit. Orbit with energy only slightly less than escape energy.

Cavity reactor. Nuclear reactor with all fissioning material in gaseous form. Gaseous core reactor.

Centrifugal acceleration. Lateral acceleration experienced by a body moving on a circular path.

Circular velocity. Velocity necessary for maintaining a circular orbit.

Conic section. Curve formed by the intersection of a plane and a right circular cone. Usually called "conic." There are three types —ellipse, parabola, and hyperbola.

Critical mass. Minimum mass required for nuclear fission.

Cryogenic. Pertaining to very low temperature at which normal gases become liquids or solids.

Density. Amount of matter per unit volume.

Deuturium. Heavy hydrogen. Hydrogen atoms which have a neutron in the nucleus as well as a proton.

Dilution ratio. Ratio of amount of propellant flowing through a nuclear reactor to the amount of nuclear fuel burned.

Dissociation. The breaking of neutral molecules into free radicals and atoms.

Drag. Retarding force experienced by an object moving through a fluid or gas.

Drag loss. Velocity lost by a vehicle moving through an atmosphere due to the drag force.

Earth-storable liquid propellants. Liquid propellants which can be easily stored at the earth's surface with little loss of propellant due to evaporation, etc.

Earth time. In relativistic calculations, the time which passes on earth as opposed to the time which passes in a starship.

Ecliptic (plane). Plane of the earth's orbit around the sun, inclined to the earth's equator by about $23°45'$.

Electrical rocket. Rocket which uses some form of electrical device such as an arc jet, ion engine, or magnetohydrodynamic accelerator to accelerate a propellant to achieve thrust.

Ellipse. One of the conic sections. The elongated circular path followed by bodies with less than escape velocity.

Energy. Ability to do work, measured in foot-pounds.

Enthalpy. Total heat content.

Escape velocity. Velocity which a particle or larger body must attain in order to escape from the gravitational field of a planet or star.

Exhaust velocity. Average velocity at which the exhaust gases are expelled from the nozzle of a rocket.

Expansion ratio. Ratio of the exit area of a nozzle to its throat area.

Fast neutron. Neutron with relatively high velocity compared to other neutrons.

Fission. The splitting apart of a large atom into two or more smaller pieces with release of large amounts of energy.

Fission products. Atoms or other particles, frequently radioactive, produced by the fission process.

Flight-path angle. Angle made by the tangent to the flight path (or trajectory) with the horizontal.

Flyby. Passing close to a planet as opposed to orbiting or landing upon it.

Force. That which changes the state of rest or motion in matter. The rate of change of momentum.

Foot-pound. Work done in moving a force of one pound parallel to itself through a distance of one foot.

Free radical. Electrically neutral atom or group of atoms with unstable electronic configuration.

Fuel. Material which is used to release energy in chemical reactions usually in combination with an oxidizer.

Fusion. Combining of two small atoms into a single larger one with release of large amounts of energy.

Galaxy. Large cluster of stars. Our galaxy, the Milky Way, contains about 100 billion stars.

Gamma ray. Very high-energy x rays. Also called gamma radiation.

Gaseous-core rocket. Nuclear rocket with all nuclear fuel in gaseous form.

Geocentric. Relative to the earth as a center. Measured from the center of the earth.

Gram. A measure of weight. 454 grams equals one pound.

Gravity. Force imparted by the earth to a mass on or close to the earth.

Gravity loss. Velocity lost due to part of the thrust on a rocket being nullified by the force of gravity.

Heliocentric. Relative to the sun as a center. Measured from the center of the sun.

Hohmann transfer. Elliptical path between two circular orbits which requires least energy expenditure.

Horsepower. Rate of expenditure of energy. One horsepower equals 550 foot-pounds per second.

Hybrid propellant. Propellant system using both a solid and a liquid propellant.

Hyperbola. One of the conic sections. The open trajectory followed by a body with more than escape velocity.

Hyperbolic excess velocity. Velocity remaining at an infinitely great distance from a gravitating body.

Hypergolic. Combinations of chemical fuels and oxidizers which spontaneously ignite when brought together.

Impulsive velocity. Velocity attained by a rocket in the absence of drag and gravity.

Infinite staging. Continuous discarding of tanks and engines as propellant is consumed.

Inner solar system. Portion of the solar system bounded by the planet Mars.

Intergalactic space. Space between the galaxies.

Isotopes. Particles of the same atomic number but with differing atomic masses.

Kilo-. Prefix meaning multiplied by one thousand.

Kiloton. Energy release approximately equal to one thousand tons of high explosives.

Kinematic. Pertaining to motion.

Kinetic energy. Energy which a body possesses by virtue of its motion.

Launch window. Interval of time during which a rocket can be launched to accomplish a particular purpose.

Lift/Drag ratio (L/D). Ratio of the lift force on a body flying through the atmosphere to its drag force.

Light-year. Distance light travels in one year. Equal to 5.9×10^{12} miles.

Linear momentum. Momentum in the direction of motion. Equal to the product of mass and velocity.

Liquid-core nuclear rocket. Nuclear rocket in which the nuclear fuel is in liquid form.

Liquid-propellant rocket (chemical). Rocket engine fueled with propellant or propellants in liquid form.

Major planets. The massive planets of the solar system—Jupiter, Saturn, Neptune, and Uranus.

Mass. Measure of the amount of matter or number of molecules in a material object or body.

Mass annihilation. Release of energy by complete annihilation of mass.

Mega-. Prefix meaning multiplied by one million.

Megaton. Energy release approximately equal to one million tons of high explosive.

Milli-. Prefix meaning divided by one thousand.

Minor planets. The small planets of the solar system—Mercury, Venus, Earth, Mars, and Pluto.

Moderator. Material which slows down neutrons.

Molecular weight. Weight of an individual molecule, usually given in multiples of the weight of a proton.

Momentum thrust. Thrust due to the recoil of the exhaust gases from a nozzle.

Multiple star. Star system consisting of two or more stars revolving around each other.

Natural satellite. Small bodies (moons) rotating around larger bodies (planets) in the solar system.

Neutron. Subatomic particle with no electrical charge, and with a mass slightly more than the mass of the proton.

Nozzle. That part of a rocket thrust unit in which gases produced in the chamber are accelerated to high velocities.

Orbit. Path of a body about a source of gravitation.

Orbital velocity. Velocity of an object in an orbit.

Out-of-ecliptic. Region of space which does not include the ecliptic plane.

Outer solar system. Portion of the solar system outside of the orbit of the planet Mars.

Oxidizer. Material which is used to release energy in chemical reactions in combination with a fuel.

Parabola. One of the conic sections. Open trajectory followed by a body with escape velocity.

Periapsis. Point on the orbit of an object which is closest to the source of gravitation.

Perigee. Point on the orbit of an object about the earth which is closest to the earth.

Perihelion. Point on the orbit of an object about the sun which is closest to the sun.

Period. Time required for an object on a closed orbit to complete one trip around the source of gravitation.

Photon rocket. Rocket which derives its thrust from a beam of photons rather than material particles.

Poison (neutron). Material which absorbs neutrons readily.

Positron. Particle similar to electron but with positive electrical charge.

Potential energy. Energy of position in a gravitational field, measured in foot-pounds.

Power. Time rate of expending energy or doing work, measured in foot-pounds per second, horsepower, or watts.

Power density. Power produced per unit volume.

Pressure. Force exerted per unit area of a surface.

Pressure thrust. Thrust due to atmospheric pressure at the exit of a nozzle.

Propellant. In chemical rockets, fuel and oxidizer used for propulsion. In nuclear rockets, the working fluid heated by the nuclear fuel and expelled for propulsion.

Radiation. Energy transmitted by electromagnetic waves rather than by particles.

Radioisotopes. Atomic particles which decay by natural radioactivity.

Reactor (nuclear). Assembly of nuclear fuel, moderators, reflectors, etc., which creates and controls a nuclear energy release.

Recombination. Process by which free radicals and atoms join to form neutral molecules.

Reflector. Material which reflects neutrons in a nuclear reactor.

Rem. Roentgen equivalent man. A measure of biological damage causing ability of nuclear radiation.

Regenerative cooling. Cooling of a rocket combustion chamber or nozzle by circulating fuel or oxidizer, or both, around the part to be cooled.

Separation ratio. Amount of unburned fuel escaping in the exhaust of a nuclear rocket compared to the amount of propellant in the exhaust.

Shielding. Material used to absorb or reflect harmful radiation or particles from vital portions of a vehicle.

Ship time. In relativistic calculations, the time which passes in a starship as opposed to the time which passes on earth.

Slug. Unit of mass weighing 32.174 pounds at the earth's surface.

Solid-core nuclear rocket. Nuclear rocket with all nuclear fuel in solid form.

Solid-propellant rocket (chemical). Rocket engine using a solid propellant.

Space radiator. High-temperature surface used to radiate energy into space.

Space-storable liquid propellants. Liquid propellants which can be easily stored in space with little loss of propellant due to evaporation, etc.

Stage (rocket). Propulsion unit of a rocket, especially one unit of a multistage rocket, including its own fuel and tanks.

Swing around. Close flyby of a planet to change vehicle trajectory.

Synodic period. Time interval between successive conjunctions of two planets.

Thermal energy. Energy a body or gas possesses by virtue of its temperature.

Thermal neutrons. Neutrons at the same temperature as the material through which they move.

Thermal reactor. Nuclear reactor which operates on thermal neutrons.

Thrust. Force exerted by the exhaust of a rocket engine.

Time dilation. Slowing down of time which occurs at velocities close to the velocity of light.

Tripropellant. In chemical rockets, the use of three propellant substances.

Undiluted specific impulse. Specific impulse of a nuclear reaction with no additional propellant added to reduce the temperature.

Vacuum thrust. Thrust of a rocket in a vacuum.

Velocity. Rate at which distance is covered.

Watt. Measure of power—746 watts equals one horsepower.

Weight. Force exerted by a gravitational field upon a mass, measured in pounds.

Work. Product of force and distance when the force is moving a mass against a resistance, measured in foot-pounds.

Index

Acceleration, centrifugal, 54; electromagnetic, 186–187; gravitational, 9–10, 29–30; rocket, 42–43, 148–152

Air-breathing engines, 137–139

Airplane, armaments, 42; energy requirements, 66–67; lift/drag ratio, 33–35; nuclear, 134–135

Alpha Centauri, 188–191

Antimatter, 203–207

Apogee, 56–57; of escape orbit, 61

Artillery rockets, 28–51

Asteroids, 106–108, 185–187

Astronomical unit, 90, 141

Atmospheric, braking, 88–89, 143–145; drag, 31–52, 54; pressure and rockets, 13, 18–20, 41

Automobile, energy requirements, 67; lift/drag ratio, 35–37

Ballistic rockets, 28–52, 59–61, 69–72

Biological time dilation, 193–195, 207

Braking, atmospheric, 88–89, 143; in gravity field, 87–88, 105

Celestial mechanics, 78–112, 140–145

Circular orbit, 53–58, 61–63

Combustion, rocket, 15–18, 39–41

Comets, 106–108

Communication, with life in space, 191–193

Conjunction, planets, 93

Cooling, rocket engine, 159–163

Cost of space travel, 66–68, 172–179, 197–201

Critical mass, 127

Cryogenic liquid propellants, 45, 51, 117–122

Drag, atmospheric, 31–32, 54; intergalactic space, 207

Earth, escape velocity, 7, 55–56, 61–63; gravitational potential, 29–30, 55, 66, 84–85

Ecliptic plane, 79

Efficiency, gun, 10–11; rocket, 15–18, 24–26, 39–41, 50–51, 187

Electrical rockets, 148–152; nuclear, 164–169

Electromagnetic gun, 185–187

Elliptical trajectory, 53, 56–58

Energy, conversions, 8–11; kinetic, 10, 25–26, 29–30; mass and, 127; for orbit, 61–63; of planet, 87, 95–102; potential, 29–30, 55, 63; solar, 169–170; for space travel, 66–68

Escape velocity, 7, 55–56, 61, 84–87; galaxy, 195–196; moon, 89–90; planets, 63–65, 83, 143; solar, 141; solar system, 7, 86, 102

Exhaust velocity, 12–13, 17–23

Fission, 127–132

Fuel, *See* Propellant

Fusion, 127–128; rockets, 196–201

Galaxy, 190–191; escape, 195–196

Gaseous-core nuclear rocket, 154–157; cost of operation, 172–179

German V-2 rocket, 28–29, 49–51

Goddard, Robert, 23, 39–44, 48–50

Gravity field, braking in, 87–88;

force of, 9–10, 29–30, 32–33, 52–55; 65–68; high-velocity rockets and, 86–87, 140–143; of planets, 78–87, 95–102

Gravity loss, 32–33, 65, 86–87, 140–143

Guns, operation of, 8–12; rockets and, 12, 31, 37, 42–43, 51

Heliocentric velocity, 85, 143

Hohmann Transfers, 61–63, 90–95

Hybrid propellant rocket, 73–74, 123

Hyperbolic, boost, 150–151; excess velocity, 84–87, 140–143, 195–196; trajectory, 78

Hypergolic propellant, 73

Impulse, *See* Specific impulse

ICBM, 7, 59–60, 74–76

Interplanetary travel, 78–102, 140–187; launch windows, 91–92, 108–112; non-rocket, 185–187; refueling bases, 112–115; space ships, 140–187; time for, 93, 103–105, 141–147, 151–152

Interstellar travel, 188–207

Isotopes, as heat source, 132–134

Kepler's laws, planets, 52–53, 56

Kinetic energy, 10; exhaust, 25–26; potential energy and, 29–30, 63

Launch window, 91, 108–112

Life, other star systems, 191–193

Lift/drag ratio, 33–37

Liquid-core nuclear rocket, 154

Low-thrust rocket, 148–153

Mass, conversion to energy, 127; critical, 127; force and, 9–10

Mass-annihilation rocket, 203–207

Minimum energy trajectory, 58, 90

Momentum, conservation of, 8, 56–58; thrust, 13

Moon, escape velocity, 89–90; re-fueling on, 112–115; velocity for travel to, 7, 89–90

Motion, energy of, 10; laws of, 8–10, 13; of planets, 52–53

Multi-body gravity fields, 78–84

Multi-impulse trajectory, 93–95

Nozzle, rocket, 8, 15–20, 124–127

Nuclear, fission, 127–132; fusion, 127–128; reactors, 134–135; war-heads, 43

Nuclear rockets, electric, 164–169; gaseous-core, 154–157, 172–173; liquid-core, 154; pulse, 170–171; radioisotope-heated, 132–134; solid-core, 128–135; thermal, 127–135, 152–163

Orbit, circular, 53–54, 61–63; elip-tical, 56–58; of planets, 52–53, 78, 80–83; transfer, 61–63, 90–95

Orbital rocket, 52–77

Orbital velocity, 7, 56–61

Parabolic trajectory, 53, 55–56

Payload, ballistic missile, 43, 75–76; electric rocket, 166–168; spaceship, 176–177

Perigee, 56–57

Period, of elliptical orbit, 58; of planets, 79–81, 92–93; of satel-lite, 54

Photon rocket, 201–203

Planets, conjunction of, 93; energy of, 95–102; escape velocity of, 63–65, 78–89, 143; flyby, 95–102; landing on, 88–89; launch window, 91–92, 109–112; orbit, 52–53, 78, 80–83; refueling bases on, 115, 145–148; in space, 188–191; synodic periods of, 92–93, 108–112; 143–145; transfer orbits between, 90–95

Plug-nozzle engine, 126–127

Potential energy, 29–30, 55, 63

Power, 11–12; pump, 20; of rocket exhaust, 14–15

Pressure, in rocket engine, 41, 47–48, 124–127; thrust and, 13, 18–20

Propellant, combustion tempera-ture, 17–18; consumption, 14, 157–159; cost, 66–68, 172–173;

cryogenic, 45, 48–51; liquid, 20, 44–51, 72-77, 115–123; requiring air, 137–139; solid, 38–43, 47–48, 68–72

Radiation, *See* Shielding

Radioisotope solid-core rocket, 132–134

Range, of ballistic rocket, 28–29, 58–61

Relativity, theory of, 193–195

Rockets, artillery, 28–51; ballistic, 28–33; cooling, 159–163; electrical, 148–152; fusion, 196–201; global, 60–77; high-velocity, 140–207; history, 38–51, 76–77; hybrid, 73–74, 123; interplanetary, 78–102; liquid-propellant, 20, 44–51, 72–77, 115–123; low-thrust, 148–152; mass annihilation, 203–207; nozzles, 8, 15–20, 39, 124–127; nuclear, 127–135, 152–171; operation of, 8–27, 50–51; orbital, 52–77; photon, 201–203; plug-nozzle, 126–127; refueling in space, 112–115; solar-powered, 169–170; solid-propellant, 38–42, 47–48, 68–72; sounding, 45–47; versatility of, 76–77

Rocket equation, 21–23, 39

Satellite, in circular orbit, 54; escape velocity of, 63–65; landing on, 105; refueling bases on, 145; synchronous, 62

Shielding, nuclear rockets, 128, 133–135; 198; space radiation, 182

Ship, energy requirements, 67–68; lift/drag ratio, 35–37

Solar-powered rockets, 169–170

Solar probes, 94–95, 100–102

Solar system, data, 79–83; escape, 7, 86, 102, 141–148; *see also* Interplanetary travel; Planets

Solid-core nuclear rockets, 128–135

Sounding rockets, 45–47

Space, communication with life, 191–193; intergalactic, 207; philosophy of exploration, 208–211; probability of planets in, 188–191

Spaceship, costs, 173–179; design, 171–187; interplanetary, 140–187; interstellar, 188–207

Space stations, 113–114

Specific impulse, 14, 18, 68, 70, 75; liquid propellants, 45–48, 72–75, 118–122; nuclear rockets, 153–158, 164, 170; solid propellants, 42–44, 47–48, 68–72; velocity and, 116–117

Stars, 188–196

Synodic period, *See* Planets

Thrust, 8, 12–20, 124–127

Thrust/weight ratio, 69, 162–166, 169–170, 198–200

Time, dilation of, 193–195

Trajectory, deflection by planet, 95–102; elliptical, 53, 56–58; general, 59–61; hyperbolic, 53, 78; minimum energy, 58–59, 90; multi-impulse, 93–95; out-of-ecliptic, 99–102; parabolic, 53, 55–56

Transfer orbits, 61–63, 90–95; minimum energy, 103 105

Useful load, of rocket, 23–24

Velocity, atmospheric drag and, 31–32; exhaust, 12–18, 21–23; gravity loss and, 32–33, 86–87, 140–143; heliocentric, 85; Hohmann Transfer, 90–95; hyperbolic excess, 84–87; initial, 26–27, 30; interplanetary travel, 140–187; interstellar travel, 188–207; of light, 193–195; lunar mission, 7, 89–90, 112; orbital, 7, 52–77; parabolic, 55–56; rocket, 7–8, 21–23, 26–27; solar probe, 100–102; specific impulse and, 138

Vis-Viva Law, 56–58, 85

Weight, 9–10; of rockets, 9, 21–23, 26–27, 47–48

2016 Addenda to *THRUST INTO SPACE*

Publications and Papers

Maxwell W. Hunter, II

The list below is in chronological order. More information is available at www.maxwellhunter.com/publications-and-papers.

"Impulsive Midcourse Correction of a Lunar Shot"
(with W.B. Kempere and R.J. Gunkel)
Douglas Aircraft Company, 1958
MHF.PD01

"Manned Nuclear Space Systems"
(with E.B. Konecci and R.F. Trapp)
Appeared in Aero Space Engineering, October 1959
Douglas Aircraft Company\
MHF.PD02

"The Advantages of High Thrust Space Vehicles"
(with J.M. Tschirgi)
Douglas Aircraft Company, November 1959
MHF.PD03

"The Potential of Nuclear Space Transport Systems"
(with W.E. Matheson and R.F. Trapp)
Douglas Aircraft Company, March 1960
MHF.PD04

"Direct Operating Cost Analysis of a Class of Nuclear Spaceships"
(with W.E. Matheson and R.F. Trapp)
11th Congress, International Astronautical Federation, Stockholm, August 1960
Douglas Aircraft Company
MHF.PD05

"The Cost of Lunar Travel"
International Congress & Exposition of Automotive Engineering, January 1961
Douglas Aircraft Company
MHF.PD06

"RITA – The Reusable Interplanetary Transport Approach"
Douglas Aircraft Company, February 1961
MHF.PD07

"Early Interplanetary Exploration with Nuclear Rockets"
(with R.F. Trapp)
Japanese Rocket Society, Tokyo, August 1961
Douglas Aircraft Company
MHF.PD08

"Advanced Manned Space Systems"
(with E.B. Konecci)
American Rocket Society, Space Flight Report to the Nation, New York City, October 1961
Douglas Aircraft Company
MHF.PD09

"Capabilities of Advanced Vehicles for Astronomical Research"
(with D.S. Merrilees)
Appeared in Space Age Astronomy August 1962
Douglas Aircraft Company
MHF.PD10

"The Potential for Nuclear Propulsion for Manned Space Flight"
National Meeting on Manned Space Flight, Institute of the Aerospace Sciences, May 1962
National Aeronautics and Space Council
MHF.PS01

"Recoverable Space Launching Nuclear Systems"
Seventh Symposium on Ballistic Missiles and Space Technology, U.S.Air Force Academy,
August 1962
National Aeronautics and Space Council
MHF.PS02

"Single-Stage Spaceships Should be our Goal"
Appeared in Nucleonics, February 1963
National Aeronautics and Space Council
MHF.PS03

"Possible Nuclear Space Vehicles During this Century"
National Topic Meeting on Nuclear Materials for Space Applications, American Nuclear
Society, Cincinnati, April 1963
National Aeronautics and Space Council
MHF.PS04

"Barriermanship" or "How to Walk a Development Program"
Washington D.C., 1963
National Aeronautics and Space Council
MHF.PS04

"Commercial Space Transportation Possibilities"
13th Annual Meeting of the American Astronautical Society, Dallas, May 1967
Lockheed Missiles & Space Company
MHF.PL04

"Unmanned Exploration of the Inner Planets, Comets, the Outer Planets, and Near Interstellar Space"
American Astronautical Society Conference on the Role of Planetary Geology and Geophysics in the Space Effort, Boston, May 1967
Lockheed Missiles & Space Company
MHF.PL05

"Nuclear Propelled Passenger Vehicle for Lunar and Planetary Exploration"
(with C.D. McKereghan)
Second Conference on Planetology and Space Mission Planning, New York Academy of Sciences, New York, October 1967
Lockheed Missiles & Space Company
MHF.PL06

"Propulsion Systems for Manned Planetary Missions"
1968 Space Technology Conference, Society of Automotive Engineers, Washington, DC May 1968
Lockheed Missiles & Space Company
MHF.PL07

"Accessible Regions Beyond the Solar System"
Joint National Meeting, American Astronautical Society/Operations Research Society, Denver, June 1969
Lockheed Missiles & Space Company
MHF.PL08

"Are Technological Upheavals Inevitable?"
Harvard Business Review, September-October 1969
Lockheed Missiles & Space Company
MHF.PL09

"The Effect of Transportation Systems on Manned Planetary Precursor Missions"
American Astronautical Society, 16th Annual Meeting, "Space Stations and Manned Planetary Exploration," Anaheim, California June 1970
Lockheed Missiles & Space Company
MHF.PL10

"The Hypersonic Transport – The Technology and the Potential"
American Institute of Aeronautics and Astronautics, 7th Annual Meeting and Technical
Display, Houston, October 1970
Lockheed Missiles & Space Company
MHF.PL11

"Space Shuttle Influence on Payload and Spacecraft Design"
22nd International Astronautical Congress, Brussels, Belgium, September 20-25, 1971
Lockheed Missiles & Space Company
MHF.PL12

"An Analysis of the Effects of the Space Shuttle on Payloads"
5th U.S.-European Conference, San Francisco, May 1972
Lockheed Missiles & Space Company
MHF.PL13

"Payload Design Concepts for the Space Shuttle"
(with R.M. Gary, W.F. Miller)
Astronautics & Aeronautics, June 1972
Lockheed Missiles & Space Company
MHF.PL14

"The Origins of the Shuttle (According to Hunter)"
Lockheed Missiles and Space Company, Inc., Sunnyvale, California, September 1972
Lockheed Missiles & Space Company
MHF.PL15

**"The Effect of Space Shuttle Payload Design Techniques on Total Space
Program Cost"**
23rd IAF Congress, Vienna, Austria, October 1972
Lockheed Missiles & Space Company
MHF.PL16

"Design of Low-Cost, Refurbishable Spacecraft for Use With the Shuttle"
(with R.M. Gray, W.F. Miller)
AIAA 9th Annual Meeting and Technical Display (AIAA Paper No. 73-73), Washington, DC,
January 1973
Lockheed Missiles & Space Company
MHF.PL17

"An Engineer's View of The Large Space Telescope (LST)"
Lockheed Missiles and Space Company, Inc,, Sunnyvale, California, April 1975
Lockheed Missiles & Space Company
MHF.PL18

"Standards Components and Modules for Space Shuttle Payloads"
UCLA Extension, Los Angeles, May 1975
Lockheed Missiles & Space Company
MHF.PL19

"A Galactic Perspective"
First Foothill College Symposium on God and Man: Our Future in Space, Los Altos Hills, California, July 1975
Lockheed Missiles & Space Company
MHF.PL20

"Beyond the Geocoronium"
A Tribute to Robert H. Goddard, Smithsonian Institute, Washington, DC, March 1976
Lockheed Missiles & Space Company
MHF.PL21

"The Modern Pursuit of Scientific Curiosity"
Lockheed Missiles and Space Company, Inc., Sunnyvale, California, April 1976
Lockheed Missiles & Space Company
MHF.PL22

"The Relativity of Relevance"
Second Annual Foothill College Symposium on God and Humanity, Los Altos Hills, California, October 2, 1976
Lockheed Missiles & Space Company
MHF.PL23

"Strategic Dynamics and Space Laser Weaponry"
Lockheed Missiles and Space, Co., Inc. Sunnyvale, California October 31, 1977
Lockheed Missiles & Space Company
MHF.PL24

"Transportation Options and High Payoff Choices"
AIAA 1982 Annual Meeting and technical Display, Baltimore, MD, May 25-27, 1982
Lockheed Missiles & Space Company
MHF.PL25

"Advanced Space Transportation Options"
NGOS at Unispace '82, Vienna, Austria, August 9-21, 1982
Lockheed Missiles & Space Company
MHF.PL26

"The Relativity of Relevance"
Second Annual Foothill College Symposium on God and Humanity
Los Altos Hills, CA, October 2, 1976
MHF.H04

"Acceptance Speech - On receiving the Outstanding Alumni Award"
Hollidaysburg High School, Hollidaysburg, PA, June 30, 1984
MHF.H05

"The Opportunity"
San Carlos, CA, August 6, 1985 - Rev. May 6, 1986
MHF.H06

"The Art of Weaponry"
San Carlos, CA, November, 1985
MHF.H07

"Great Zeus!"
San Carlos, CA, July 4, 1987
MHF.H08

"Four Easy Correlations"
San Carlos, CA, August 7, 1987
MHF.H09

"The Battle"
San Carlos, CA, 1986
MHF.H10

"The SSX - Single Stage Experimental Rocket"
San Carlos, CA, March 15, 1988
MHF.H11

"The SSX - Designing for Flight Safety"
San Carlos, CA, December 18, 1988
MHF.H12

"SSX - Hydrogen vs Hydrocarbons"
San Carlos, CA, December 18, 1988
MHF.H13

"A Reusable Commercial Space Transport in a World of Expendables: Development, Certification and Operation"
with Gary C. Hudson, Pacific American Launch Systems, Menlo Park, CA, April, 1989
MHF.H14

"The SSX - Space Command, Space Force and Spaceships"
San Carlos, CA, Sept 15, 1989
MHF.H15

"The SSX - A True SpaceShip"
The Journal of Practical Applications in Space, Arlington, VA, Fall 1989
MHF.H16

"The SSX - SpaceShip Experimental - Seven Years Later"
Presented to the 40th annual meeting of the Japanese Rocket Society, Tokyo, Japan, 22 April 1996
MHF.H17

"The SpaceGuild Plan"
San Carlos, CA May 17, 1989
MHF.G01

"SpaceGuild Related Activities"
San Carlos, CA May 17, 1989
MHF.G02

"The Rationale for SpaceGuild"
San Carlos, CA Feb 12, 1990
MHF.G03

Supplemental Biography*
Maxwell W. Hunter II

Maxwell W. Hunter II, a visionary rocketeer, consummate aerodynamicist, project manager and policy adviser for five decades, steadfastly contributed to the nation's rocket programs and the future of space transportation.

After graduating from MIT in 1944, Max began his career at Douglas Aircraft Company in Santa Monica, California. During his 18-year tenure there, he was responsible for the aerodynamic design of Nike-Ajax and Hercules, Sparrows I, II and III, Honest John and other missiles. He was later, as Chief Missiles Design Engineer, responsible for the design of Thor, Nike-Zeus and others and as Chief Engineer of Space Systems, for the engineering of all Douglas space efforts including the Delta, the Saturn S-IV and S-IVB stages and others.

In 1962, Max joined the professional staff of the National Aeronautics and Space Council in Washington D.C. As part of this advisory group to the President of the United States, he provided insight into future space programs and the creation of national space policy in both the Kennedy and Johnson administrations. While there, he was the first to recognize the strong effect of Jupiter's gravity on planetary probe vehicles and was instrumental in opening up the outer solar system by supplementing rocket performance with planetary gravitational impulse.

Max was with the Lockheed Missiles and Space Company for 22 years from 1965 to 1987, where he was responsible for the design of

* The original Max Hunter biography, current at the time of publication, can be found at the front of this book.

the advanced space transportation vehicles StarClipper and Shuttle and originated the concept of using large expendable tanks in shuttle design. He led the team that won the contract for the design of the large space telescope (eventually dubbed Hubble Space Telescope) and served as program manager for four years during its creation phase.

Max did extensive work in the applications of high energy lasers for missile defense and originated the space laser battle station concept. His 1977 paper "Strategic Dynamics and Space Laser Weaponry" has often been cited as starting the conversation that resulted in President Reagan's Strategic Defense Initiative (SDI).

As founder and president of his consulting firm, SpaceGuild, Max was instrumental in starting the SSTO (Single-Stage-to Orbit) program in 1990 and worked closely with the McDonnell Douglas Delta Clipper team. The Delta Clipper DC-X first flew on August 18, 1993 at White Sands Missile Range, New Mexico.

For over 35 years, starting in 1959 with the RITA single stage nuclear rocket at Douglas, the StarClipper expendable tank design at Lockheed in 1966 and the X-Rocket single stage to orbit at Lockheed in 1985, he continually pursued single stage designs and aircraft-like operations as the keys to vastly improved, economical space transportation. The potential of this approach was dramatically demonstrated in 1993 with the test flights of the Delta Clipper.

He authored over five dozen technical papers. Their subject matter has included the unmanned exploration of the solar system and the economics of manned space transportation, the latter dealing with both utilization of advanced nuclear rockets and the use of chemical rockets and expendable tanks in space shuttles. He has also authored "Are Technological Upheavals Inevitable?" published in the Harvard Business Review and a rocket propulsion textbook titled "Thrust into Space."

Mr. Hunter graduated from Hollidaysburg High School, Hollidaysburg, Pennsylvania in 1939. He received an A.B. degree in Physics and Mathematics from Washington and Jefferson College in 1942 and an M.S. degree in Aeronautical Engineers from the Massachusetts Institute of Technology in 1944. He attended the Advanced Management Program of the Harvard Business School in 1967.

He was Phi Beta Kappa, Tau Beta Pi, and a Fellow of the American Institute of Aeronautics and Astronautics, American Astronomical Society, and the British Interplanetary Society, was a member or the International Academy of Astronautics and an honorary member of the Japanese Rocket Society. In 1982 he received the NASA Distinguished Public Service medal for "the definition and promotion of the space shuttle and its utilization." In 1995 he received the Wernher von Braun Memorial Award of the National Space Society for "lifelong contributions to the fields of rockets, missiles a spaceflight."

He had five children (Max III, Sally, Peggy, Matt and Dave) with his first wife, Nancy Spencer Hunter and was married to Warner Bro's musical film star Irene Manning for 37 years.

Max Hunter passed away at Stanford Hospital in 2001.

This appears to be a press conference related to Saturn S-IV stage. (Left to right): **Maxwell W. Hunter II,** Chief Engineer of Space Systems, Douglas Aircraft Company; **Robert W. Kamm,** Director of Western Operations Office, NASA; **Wernher Von Braun**, Director Marshall Space Flight Center, NASA; **Elmer P. Wheaton,** Vice President Engineering/Technical, Douglas Aircraft Company; **Oswald H. Lange**, Mgr, Saturn Systems Office NASA, Marshall Space Flight Center.

Vice President Lyndon B. Johnson (center) presides over a meeting of the National Aeronautics and Space Council (**Maxwell W. Hunter**, lower left).

(Right) **Max Hunter** stands in front of McDonnell-Douglas' DC-X, short for Delta Clipper or Delta Clipper Experimental, an unmanned prototype of a reusable single-stage-to-orbit launch vehichle built in conjunction with the United States Dept of Defense's Strategic Defense Initiative Organization (SDIO) from 1991 to 1993.

(Left) During his tenure at Douglas Aircraft, **Max** was responsible for the design of the Nike-Ajax and Nike-Hercules anti-aircraft missiles and the Nike-Zeus anti-ballistic missile.

Maxwell W. Hunter Foundation

The non-profit, educational Maxwell W. Hunter Foundation's mission is to inspire a new generation of space enthusiasts. The Foundation works to:

1. Advocate for the relentless advancement of space exploration and innovation in space transportation in the tradition of Max Hunter's work;
2. Support secondary-level aeronautical engineering curriculum to inspire the boundless imagination of America's youth;
3. Provide insight into America's golden age of space exploration from Max Hunter's unique perspective, through his personal library of papers and articles he kept during his five-decade career.

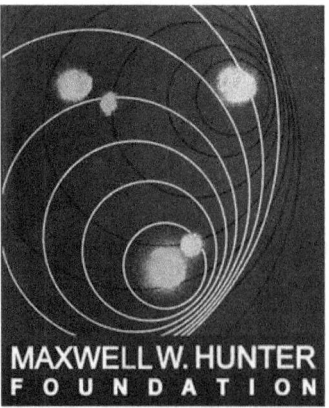

Visit MaxwellHunter.com

The official website of Max Hunter and the Maxwell W. Hunter Foundation, *MaxwellHunter.com* is a great resource for additional information on the career of Max Hunter including a history of his major projects, dozens of his technical publications and policy papers and personal musings as he reflects on America's golden age of space exploration.

Acknowledgements

This official reprint of *Thrust Into Space* was made possible by the contributions, support and enthusiasm of the following:

Dr. J.D. Crouch, II

David Hunter

Matt Hunter

Hunter Family Trust

Maxwell W. Hunter Foundation

Nancy Spencer Hunter

Nathan Hunter

Pat Hunter

Peggy Norman

Jerry Papazian

Dr. Mark Sumner

Joseph Vella

Len Whitney

Sally Wiley

Plus… The many friends, colleagues and associates of Max Hunter who have urged that *Thrust Into Space* be reprinted and made available to a new generation of space enthusiasts pursuing an education in space, aerodynamics and related engineering fields.